D1201599

INTRODUCTION TO MAGNETIC RESONANCE

Introduction to

Modern Physics Monograph Series

EDITOR: FELIX VILLARS,

Massachusetts Institute of Technology

ROBERT T. SCHUMACHER,

Carnegie-Mellon University

Introduction to Magnetic Resonance:

Principles and Applications

Magnetic Resonance

Principles and Applications

ROBERT T. SCHUMACHER

Carnegie-Mellon University

W. A. Benjamin, Inc. · New York, 1970

INTRODUCTION TO MAGNETIC RESONANCE

Principles and Applications

Standard Book Number 8053-8504-5 (Clothbound Edition)
8053-8505-3 (Paperback Edition)
Library of Congress Catalog Card Number 70-102267
Manufactured in the United States of America
12345K43210

W. A. BENJAMIN, INC.
New York, New York 10016

Editor's Foreword

Undergraduate teaching in physics is going through a phase of rapid evolution. On the frontier of physics new information is literally pouring in, new perspectives are opening up, and new concepts emerging. For the beginning student, the distance to be covered from the freshman year to graduate research work is constantly expanding.

Professional education in physics must therefore deal with the very real problem of the need for thoughtful condensation of the material presented, and the question of what may and should reasonably be achieved in four years of undergraduate instruction.

It is generally agreed, on the one hand, that a thorough presentation of the fundamentals of both classical and elementary quantum physics is essential. On the other hand, it is understandably desirable to involve the student in the excitement offered by the many interesting new developments in all fields of physics. The discussion of such new topics would provide the student with an opportunity to observe the actual growth process of science: new experiments, new techniques, and the attempts to relate new results to existing or emerging theoretical views. The study of the well-established, introductory subjects of physics appears to lack these exciting aspects and to offer little room for the display of creativity, except as a historical fact.

It has at last been recognized that this need not entirely be so; that in fact the close ties between the traditional and the modern can be exploited to establish relationships between the classical subjects and current endeavors (e.g., classical mechanics and space navigation, wave optics and radar interferometry, etc.). To develop such links wherever they exist and to put the essential parts of the traditional subjects in a modern perspective is an urgent and rewarding task. There is clearly no general agreement about the manner in which a broad subject can be reduced to its essentials, or what these essentials are, in the context of undergraduate education. The great variety of educational situations will naturally give rise to a diversity of approaches. Interested teachers will find their own challenge and excitement in experimenting with various alternatives. The recent burgeoning of such introductory undergraudate texts as the Feynman lectures, the Berkeley physics course, and the Massachusetts Institute of Technology introductory physics texts bear witness to the interest that has been aroused by the problem of bringing the fundamentals of physics to the undergraduate in a novel way.

Therefore, a new series entitled *Modern Physics Monographs* has been undertaken which intends to continue this process. This series will present material for the post-introductory undergraduate courses, that is for those normally given in the junior and senior years. At this level, there exist on the one hand courses of a specialized nature, which offer the undergraduate student an introduction to the great diversity of topics of physical science, ranging from particle physics to nuclear, atomic, solid state, plasma, and astrophysics, to name but a few. On the other hand, in these two years, basic subject matter previously touched upon may be deepened and extended; indeed, many curricula carry such a "second round" covering the fundamentals again at a more advanced level.

There is an infinite variety of ways to organize this more advanced and specialized material. We hope that the *Modern Physics Monographs* series will help to provide the lecturer in the field with additional flexibility in choosing his course material and, if he is inclined to experiment, enable him to bring into his course topics not generally covered in standard textbooks. In addition, the student will have access to a variety of inexpensive collateral reading material.

For these very reasons, the books of this series are *not* intended to be textbooks, but rather monographs, that is, works that cover a more restricted area in a space of approximately 100 to 250 pages. They contain problems with and without answers, and could either supplement existing texts or be used in groups as a replacement for a single text.

The editor is aware of the fact that the presentation of a modern physics topic on an elementary level is a major didactic challenge for the writer, and so all attention is being given to publish texts that will be readable, informative, and helpful to the student. Critical comments from all interested parties are invited, and suggestions for additional texts of general interest will be welcomed.

Felix Villars

Cambridge, Massachusetts
April 1970

Preface

It is a very common and educationally healthy practice for an undergraduate physicist to do some work in a research laboratory or an advanced teaching laboratory of "research grade." Today, in contrast to twenty years ago, even the student at relatively small colleges and universities participates in research experiments or in experiments only slightly removed (in time and difficulty) from active research. The largest problem facing such students is often not just to find suitable literature to introduce them to the work of the laboratory, but to perceive what relationship a particular experimental activity has with other experiments, with other branches of physics, other branches of science, and most important of all to the student, with the concepts, information, and techniques taught in formal course work. This book is intended to serve as a supplementary text in formal courses on the undergraduate level in atomic physics, solid state physics, and even nuclear physics. It should be regarded as collateral reading material for some experiments in advanced undergraduate laboratories, and for students with summer jobs or informal laboratory "courses" in active research labs. It may also be regarded as an interim text for the beginning graduate student who might not be quite conversant enough in quantum mechanics to manage the introductory graduate level texts, such as the ones by Pake and by Slichter.

I have written from the bias of a "magnetic resonator" working in solid state physics, and as such intended the approach to the material in Chapter 3 to represent the style or point of view that will aid the student in acquiring a useful grasp of the diverse applications of magnetic resonance. In general no command of the formal apparatus of quantum mechanics is required of the reader, although I have not hesitated to introduce some quantum mechanical explanations in parallel with classical or semiclassical ideas when it seemed appropriate. However, for the most part I have tried to exploit the classical concepts of frequency modulation and random walk as introduced in Chapter 3, and have been to some extent influenced in my choice of subjects by the possibility of their explanation in terms of these concepts.

There was another reason for leaning so heavily on this conceptual crutch. The research laboratory engaged in any area of activity acquires a cryptic language of its own, which is spoken in the laboratory and at blackboard discussion but which seldom receives, these days, an imprimatur of respectability by surviving to the journal article stage. By then the ideas have become fully dressed in the elegance and precision of theoretical physics, and consequently they are hard for the student to find. In my judgment, the most characteristic qualitative idea introduced by applications of magnetic resonance is the concept of motional narrowing, and therefore I have made the complementary approaches of random walks and frequency modulation central to the book. I have also digressed whenever it seemed appropriate to indicate the application of these ideas in other areas of physics.

Somewhat apologetically I must call attention to the absence of any explicit treatment of the vast and important field of electron paramagnetic resonance in solids. I simply found that with the exception of exchange narrowing, which *is* discussed, I could not write on such subjects as g shifts, crystal field splittings, and spin Hamiltonians in the same context and with the same language that I believe unifies the rest of the book. In particular, I could not think of any way to discuss the subjects without using much more quantum mechanics than I thought appropriate. That may be the fault of my own background, since I learned whatever I do know about EPR *after* acquiring the standard skills in quantum mechanics, rather than before, as was the case when I was first introduced to NMR. For electron paramagnetic resonance, I can think of no way to simplify or clarify the introductions to the subject by Pake and by Slichter, so I have chosen to omit it entirely.

I hope students with a variety of objectives will be able to use this book without necessarily reading all of it. Chapter 1 is basic to everything, but it duplicates material which often appears in classical mechanics courses, and which should appear in all atomic or modern physics courses. Those interested in magnetic resonances in excited states may proceed directly to the relevant parts of Chapter 6. Those interested in optical pumping should get the main ideas of Chapters 2 and 3 in mind before tackling Chapter 6. (And then such students should proceed to the excellent collection of important papers assembled in a reprint volume by Bernheim.) For students whose interest is in applications to solid state physics, the first four chapters are all relevant. Chapter 5 is in its entirety an illustrative example, an attempt to mashall a large part of the previous material to describe the results of fifteen years of magnetic research on simple metals, mainly the alkalis.

There are a few problems at the end of each chapter. Most are easy and quantitative in nature, chosen to advance my opinion that quantitative as well as qualitative understanding is essential. A few problems serve the time-honored role of introducing new material the author believed interesting or important, but could not gracefully fit into the text. In most cases, I have chosen references by applying the criterion that the student should be directed to the next most accessible place he should look to find something rather than to the most complete discussion, or to the original work. I did not make a special effort to avoid reference to the original work, however, particularly when it is readable. A brief paragraph or two at the end of each chapter has been written with the above standard in mind: namely, that the reference given should be the next place for the student to go.

I think it will be obvious to any physicist who glances at this book and knows Professor Charles P. Slichter that it owes a great debt to him. I should also acknowledge that I learned about the usefulness of the explicit application of the language of frequency modulation from Professor Hans Dehmelt. The colleagues and students who contributed to my education in the nearly fifteen years since my association with Professors Slichter and Dehmelt are so numerous they can only be thanked *en masse*. Mrs. Kate Ellis and Mrs. Lillian Horton typed a draft and the final manuscript with great dispatch and unfailing cheerfulness. The production of the book was kept on schedule with the cooperation of Miss Nancy Ann Chinchor, who lightened the tedious task of reading the galley proofs.

Robert T. Schumacher

Pittsburgh, Pennsylvania
April 1970

Contents

INTRODUCTION TO MAGNETIC RESONANCE

CHAPTER 1

Basic Principles

Historically, experimental investigations into the quantum properties of angular momentum and magnetic moments followed the same course that now seems to be the most natural in introducing the subject conceptually. This first chapter is concerned with the concepts and the experiments on isolated atomic systems with angular momentum, which began with the molecular beam experiments of Stern in the 1920's and which lead naturally into the magnetic resonance experiments of Rabi in the 1930's. The material is probably familiar to all students with the background of an introductory course in modern physics. However, it is recommended that even students with confidence in their command of the subject study the chapter, if only to identify special terminology and points of view relied upon in later chapters. The student who finds the quantum mechanical references of Section 1-4 somewhat obscure should repair to the brief Appendix for some help, at least in the mathematical manipulations of the quantum mechanics of the spin $\frac{1}{2}$ system in magnetic fields.

1-1. DEFINITIONS

A system consisting of a mass undergoing circular motion about a fixed point in a plane has *angular momentum*. If the mass carries electrical charge, it has a *magnetic moment* that is proportional to the angular momentum. It is comforting to know that such simple statements are true in general for quantum mechanical systems, and that, for *magnetic dipole* moments, the proportionality factor is a scalar. The theorem stating this is an application of a powerful and ubiquitous statement known as the Wigner–Eckart theorem. We are concerned with angular momenta of various atomic, nuclear, and elementary particle systems. Table 1-1 shows the conventional symbols used for most of the systems in which we are interested. When the discussion is about an abstract angular momentum vector, we usually use the vector symbol **J**, which serves also as the total angular momentum of an atom.

Table 1-1
Conventional Symbols for Angular Momenta

System	Symbol
single electron spin	$\mathbf{S}$
electron orbit	$\mathbf{L}$
atom	$\mathbf{J} = (\mathbf{L} + \mathbf{S})$
nucleus	$\mathbf{I}$
atom including nucleus	$\mathbf{F} = \mathbf{I} + \mathbf{J}$

We also find it convenient to consider the angular momentum vector symbols to be dimensionless, and to display the units in which angular momentum is measured explicitly. The fundamental unit is, of course, $h/2\pi = \hbar$, Planck's constant. Thus, the Wigner–Eckart theorem states simply

$$\boldsymbol{\mu} = \gamma\hbar\mathbf{J} \tag{1-1}$$

where γ, the *gyromagnetic ratio* (more rationally, the magnetogyric ratio) is the scalar promised by the theorem. Now, if one pursues the example of the opening paragraph, the factor γ can be calculated immediately. The angular momentum is $|\hbar\mathbf{J}| = |\mathbf{r} \times m\mathbf{v}| = mr^2\omega$, where $\mathbf{r}$ is the orbit's radius, ω the angular frequency, m the mass, and $\mathbf{v}$ the velocity. The magnetic moment, in Gaussian units, is $\boldsymbol{\mu} = i\mathbf{A}/c$, where $A = \pi r^2$ is the orbit's area; the vector is perpendicular to the orbital plane, as in the case of the angular momentum. Thus,

$$\frac{iA}{c} = \frac{i\pi r^2}{c} = \frac{q\omega r^2}{2c} = \frac{J\hbar q}{2mc} \tag{1-2}$$

If the particle is an electron with charge $e = -4.8 \times 10^{-10}$ esu, and mass $m = 9.1 \times 10^{-28}$ g, the gyromagnetic ratio, $\gamma = e/2mc$, is related to the Bohr magneton:

$$\beta_0 = \frac{e\hbar}{2mc} = \hbar\gamma = -0.927 \times 10^{-20} \text{ ergs/G}$$

For nuclei, it is convenient to define a nuclear Bohr magneton

$$\mu_0 = \frac{|e|\,\hbar}{2Mc} = 5.05 \times 10^{-24} \text{ ergs/G}$$

where M is the proton mass.

Electrons, protons, neutrons, and μ mesons have intrinsic angular momentum. Atoms and nuclei of interest to us are compound systems, the total angular momentum and magnetic moment of which are still proportional by the Wigner–Eckart theorem, but the proportionality factor of which depends on the details of the system. Those details are conventionally absorbed into a *g factor*, or *spectroscopic splitting factor*. This factor is g_J, the Landé g factor for atoms, or $g = 2.000\ldots$, according to the Dirac equation, for electrons and μ mesons. We define the nuclear g factor by analogy. For the most part, it remains an experimental parameter characterizing nuclear moments, since, in most cases, nuclear theory is not yet able to provide better than rough estimates of its magnitude. In general, then, the expression

$$\boldsymbol{\mu}_J = g_J\left(\frac{e}{2mc}\right)\hbar\mathbf{J} = g\beta_0\,\mathbf{J} = \gamma_e\,\hbar\mathbf{J}$$

or (1-3)

$$\boldsymbol{\mu}_I = g_I\,\mu_0\,\mathbf{I} = \gamma_n\,\hbar\mathbf{I}$$

gives the relation between $\boldsymbol{\mu}$ and $\mathbf{J}$, or $\mathbf{I}$, and it defines g. Equation (1-3) also defines the gyromagnetic ratio γ, which is now $g(e/2mc)$ for a system with intrinsic angular momentum (spin). Tables, particularly the most commonly encountered tables of nuclear moments, publish a quantity called "the magnetic moment in units of the Nuclear Bohr magneton." The maximum projection of $\mathbf{J}$ along any axis occurs for the state $M_J = J$. The magnetic moment is $\boldsymbol{\mu}_J = g\beta_0\,\mathbf{J}$, and the published number is gJ.

1-2. ENERGY IN AN EXTERNAL MAGNETIC FIELD: SPATIAL QUANTIZATION

The energy E of a magnetic moment μ in an external field[1] $\mathbf{H}$ is given by the familiar expression

$$E = -\boldsymbol{\mu}\cdot\mathbf{H} \tag{1-4}$$

or, in terms of the angular momentum,

$$E = -\,g\beta_0\,Hm_J \tag{1-5}$$

[1] In a vacuum it does not matter whether one uses H or B for the magnetic field if the quantities are expressed in Gaussian units. Strictly speaking, one should use B, but conventionally most of the literature uses H, a practice that we follow. Occasionally it is important to make the distinction in solid state physics applications, and then it is well established that the *correctly calculated B* is to be used.

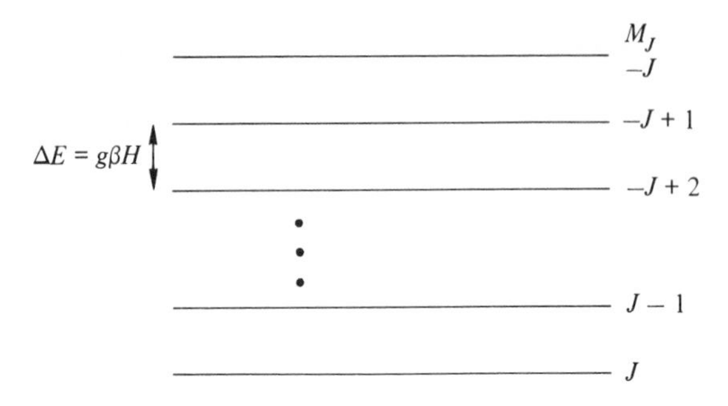

Fig. 1-1 Energy-level diagram for spin of angular momentum **J** in magnetic field **H**.

where m_J is the projection of **J** on **H**. Quantum mechanics restricts m_J to the $2J + 1$ integral or half-integral values. The energy-level diagram corresponding to Eq. (1-5) is shown in Fig. 1-1.

1-3. STERN–GERLACH EXPERIMENT

The application of a magnetic field H removes the $2J + 1$ degeneracy of the magnetic sublevels, as we have seen. Although these states are no longer degenerate, the energy differences between them are very small. In a field of 10^4 G, Eq. (1-5) corresponds to an energy separation of 1 cm^{-1} or about 10^{-4} eV for electron moments, 10^{-4} cm^{-1} or 10^{-8} eV for nuclear moments. This energy difference must be perceived against a background of 200 cm^{-1} or 0.025 eV of thermal energy at room temperature and several electron volts of energy for atomic transitions. Until the early 1920's, the consequences of spatial quantization had been manifested primarily through the Zeeman effect and the Faraday and other magneto-optic effects. The Zeeman effect was incompletely understood prior to the discovery of electron spin, and the quantitative relation of spatial quantization to the Faraday effect was obscure.

The reality of spatial quantization was demonstrated in a particularly graphic fashion by the Stern–Gerlach experiment, successfully performed in 1922. If a beam of neutral atoms passes through a homogeneous magnetic field, it is undeflected by that field, even though the magnetic degeneracy is lifted. But if the field is not spatially homogeneous, there is a net force on the moments in the beam that is given by the expression

$$\mathbf{F} = (\boldsymbol{\mu} \cdot \nabla)\mathbf{H} = \mu_x \frac{\partial \mathbf{H}}{\partial x} + \mu_y \frac{\partial \mathbf{H}}{\partial y} + \mu_z \frac{\partial \mathbf{H}}{\partial z} \tag{1-6}$$

There is a component of this force that is constant while the moment is in the gradient, and it produces a deflection of the beam that is proportional to μ_z. To see that, choose the following simplest possible field gradient. See Fig. 1-2. The beam travels in the x direction with the

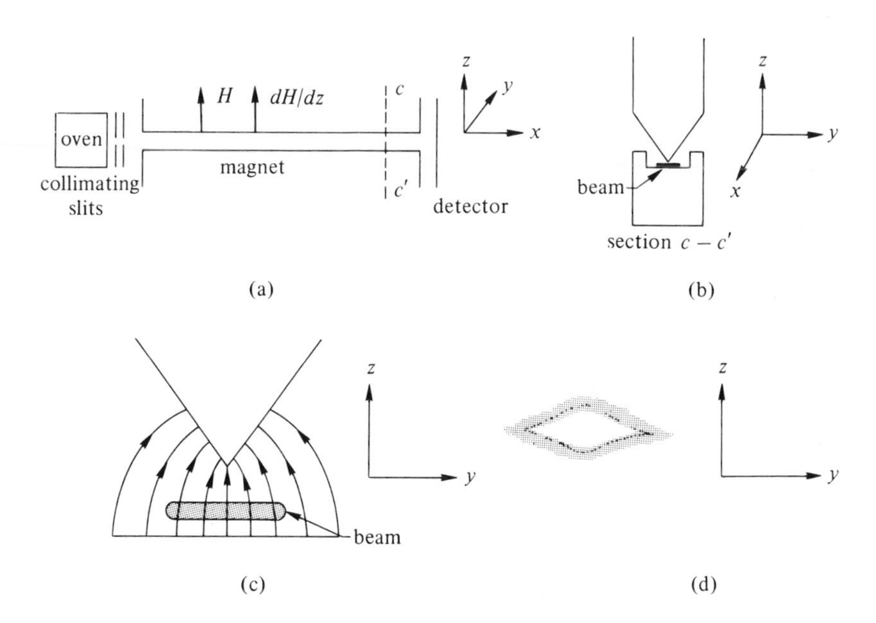

Fig. 1-2 Schematic representation of Stern–Gerlach apparatus. (a) Arrangement of main components: oven, collimating slits, magnet, and detector. (b) Cross section $c - c'$ of (a). (c) Enlarged view of beam and magnetic field in region of the beam. (d) Appearance of film deposited on substrate in original experiments of Stern.

field arranged so that $H_x = 0$. The field is principally in the z direction. All derivatives of H with respect to x vanish, and, in the beam region, both $\nabla \cdot \mathbf{H} = 0$ and $\nabla \times \mathbf{H} = 0$ are satisfied. The components of Eq. (1-6) are

$$F_x = 0 \qquad F_y = \mu_y \frac{\partial H_y}{\partial y} + \mu_z \frac{\partial H_y}{\partial z} \qquad F_z = \mu_y \frac{\partial H_z}{\partial y} + \mu_z \frac{\partial H_z}{\partial z} \tag{1-7}$$

Furthermore, let the beam lie in the symmetry plane $y = 0$, where $H_y = 0$ (Fig. 1-2c). Then $\partial H_y/\partial z = 0$, and, since $\nabla \times \mathbf{H} = 0$, $\partial H_z/\partial y = \partial H_y/\partial z = 0$. Since $\nabla \cdot \mathbf{H} = 0$, $\partial H_y/\partial y = -\partial H_z/\partial z$, Eq. (1-7) reduces to

$$F_x = 0 \qquad F_y = -\mu_y \frac{\partial H_z}{\partial z} \qquad F_z = \mu_z \frac{\partial H_z}{\partial z} \tag{1-8}$$

In the next section, we emphasize in great detail that the magnitude of the field H_z produces a torque $\boldsymbol{\mu} \times \mathbf{H}$, causing a precession about $\mathbf{H}$ (which is virtually entirely H_z at the beam coordinate) such that μ_z is constant and μ_y oscillates about an average value of zero. So the only component of the force that produces a net deflection is in the z direction, and it may be written in terms of the magnetic quantum number as $F_z = m_J g\beta_0(\partial H_z/\partial z)$. The presence of m_J means that the beam splits into $2J + 1$ components. The first experiment was done on silver (partly because the deposit could be easily "developed"), and two components were seen, as illustrated in Fig. 1-2d.[2] We now know that $2J + 1 = 2$ requires $J = \frac{1}{2}$, and that the ground state of the silver atom is an orbital S state with a single electron of spin $\frac{1}{2}$. The first experiment was done prior to the discovery of electron spin, but the result was not interpreted as requiring half-integral spin since it was assumed silver had an orbital angular momentum $L = 1$ and the $m_L = 0$ state was not allowed in old quantum theory.

Even in its simplest form the experiment has several components, none trivial, so that molecular beam experiments have long been known as the most difficult in atomic physics. The experimental problems to be solved include a high enough vacuum so that a typical beam atom can traverse the apparatus without colliding with a residual gas molecule in a meter or more of flight. The source, usually an oven with a small hole, must produce a well-collimated beam. The field gradient must be as large as possible, but the magnetic field itself cannot change too abruptly in time as sensed by the moving magnetic moment that passes from the fieldfree region to a region of maximum field gradient, and then out again to a fieldfree region before striking the detector. Finally, some device must detect, with considerable spatial resolution, the beam intensity. The modern solution of these problems is discussed in detail in the definitive monograph on molecular beams by Ramsey [2]. A somewhat briefer discussion appears in another standard reference in the field of magnetic resonance, Kopfermann's *Nuclear Moments* [3]. Among the refinements particularly useful when two Stern–Gerlach apparatuses are put in series for the standard molecular beam resonance experiment (Section 1-4) have been velocity selectors between the oven and the field region so all molecules in the beam receive the same deflection. Sophisticated universal detectors, which partially ionize the beam and send it through a simple mass spectrometer before it registers on the ultimate detector, have also

[2] The student will find it an amusing exercise in the propagation of errors to watch for illustrations such as Fig. 1-2 in which the beam is traveling in the y direction, transverse to the long dimension of the apparatus. As far as I can determine, the first such incorrect illustration appeared in A. Sommerfeld's *Atombau und Spectrallinien* [1], in all editions subsequent to 1923. It is reproduced in many texts of that school, but it also still appears in texts published in the United States as recently as 1967.

been developed. (See references listed at the end of the chapter for more discussion of experimental techniques.)

The Stern–Gerlach technique, by itself, reached its pinnacle of usefulness under the direction of Stern, particularly with the aid of Otto Frisch and I. Estermann. Although the resonance method of Rabi did prove to offer unheard of precision compared to the nonresonant experiments, the basic technique did provide a few triumphs beyond the first demonstration of spatial quantization. One of these was the discovery of the anomalous g factor of the proton (i.e., that $g_p = 5.59\ldots$ rather than $g = 2.000\ldots$, as expected from the Dirac theory of a spin $\frac{1}{2}$ particle). The initial report of this work appeared in *Nature* in 1933 [4], and it is a model of elegant brevity. The student who understands it, sentence by sentence, has a good working grasp of many of the necessary fundamentals of modern physics.

It should be emphasized that the Stern–Gerlach apparatus is a very useful practical example of a quantum mechanical state selector, or beam polarizer. The separated beams of moments of that energy from the field gradient region, each characterized by its own m_J, are polarized. The apparatus may be reversed in function and a partially or fully polarized beam sent in. Its trajectory in the apparatus is determined by the state function (i.e., the m_J level) of the constituents of the beam, so the apparatus now functions as an analyzer. These functions are important to understand and distinguish in following the magnetic resonance experiment of Rabi. A comprehensive discussion is given in volume 3 of the *Feynman Lectures on Physics* [5].

1-4. THE RABI MAGNETIC RESONANCE EXPERIMENT

If a Stern–Gerlach apparatus can be a state selector, it can also be an analyzer. What could be more natural than to put two of them in series, with some experiment in between? In the 1930's, Rabi, who had done postdoctoral work under Stern at Hamburg, performed the first magnetic resonance experiment and made the first precision nuclear magnetic moment measurement, in a homogeneous magnetic field between a polarizer and analyzer. Figure 1-3 shows two inhomogeneous fields, produced by the conventionally designated A and B magnets, with the homogeneous C magnet between them. In the C region, the magnetic resonance experiment causes transitions between magnetic quantum levels. Consider a $J = \frac{1}{2}$ system. Figure 1-3 shows the polarizer and analyzer with field gradients in the same direction. Also, care is taken that the direction of the field H itself always points in the same direction. At the end of the polarizer, one of the two separated beams may be deflected or stopped by a baffle, leaving a beam of pure $m_J = \frac{1}{2}$ particles, for instance,

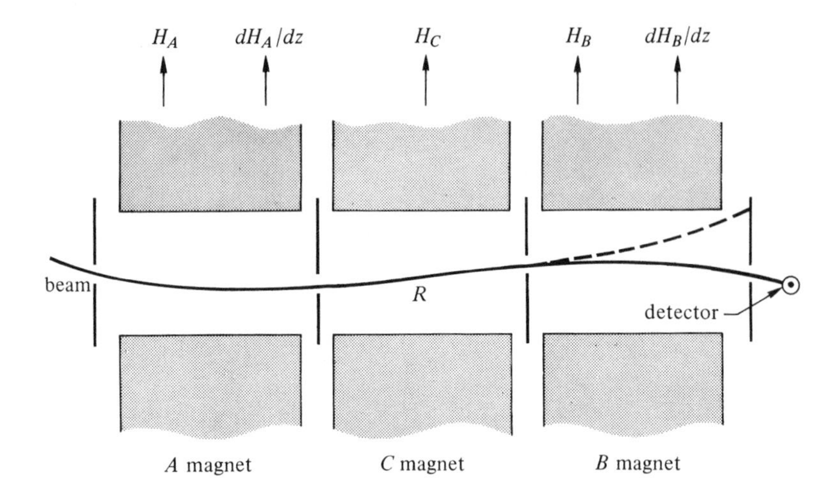

Fig. 1-3 Fields and beam trajectories in Rabi resonance experiment. The solid curved line is a greatly exaggerated trajectory for a spin $\frac{1}{2}$ system that undergoes a transition from $m_J = \frac{1}{2}$ to $m_J = -\frac{1}{2}$ in the resonance region R in the homogeneous C magnet. The dotted line in the B magnet represents the path of a molecule that does not undergo a transition and is prevented by the baffle from reaching the detector.

to enter the C magnet. If nothing is done to them there, they enter the second Stern–Gerlach apparatus where they are further deflected. If the detector is placed to detect a beam that comes through undeviated, it will detect no signal. If the beam in the C region is manipulated so that some or all of the moments are put into the $m_J = -\frac{1}{2}$ state, then in the B field their deflection will be down, and, if the A and B magnets are identical, then the B magnet will reverse the deflection produced by the A magnet, and the beam will hit the detector.

Some of the preceding details are arbitrary and of no particular importance. The experiment as described is known in the molecular beam trade as a "flop-in" experiment: the change of state in the C region causes the beam to "flop-in" to the detector. A different position of the detector, or reversal of the gradient in the B region, could result in a "flop-out" experiment. What is not unimportant is the maintenance of a magnetic field oriented in the same direction through the apparatus, or, if the field does reorient, it must do so slowly, as seen by the moment as it moves through the various regions. The first restriction is required so that the beam remains in the same quantum mechanical state unless a transition to another state is deliberately produced in the C region. The second restriction is required so that no "accidental" transitions occur between magnets.

We turn now to the theory of the magnetic resonance experiment. The quantum mechanical and classical equations of motion of a spin in a magnetic field are identical: the latter equations are for classical angular momenta and magnetic moments; the former are for expectation values of angular momentum operators. It is often sufficient to consider only the classical case. Figure 1-4 shows an angular momentum **J**, moment **μ**,

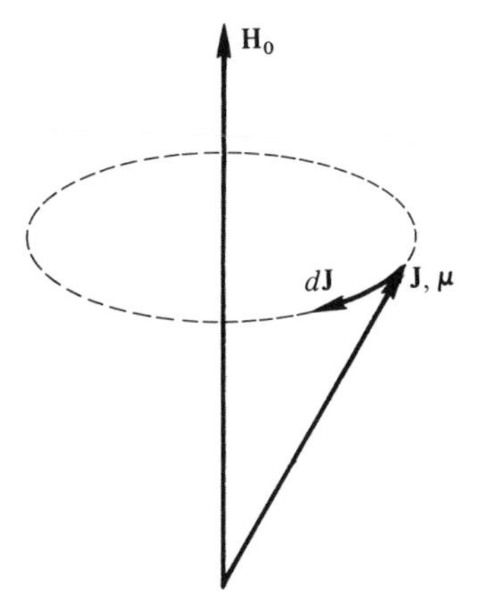

Fig. 1-4 Magnetic moment $\boldsymbol{\mu} = \gamma\hbar\mathbf{J}$ precessing in a constant field $\mathbf{H}_0$. The figure is drawn for positive γ.

inclined at an arbitrary angle with respect to the z axis, the field direction. The classical equation of motion is

$$\hbar \frac{d\mathbf{J}}{dt} = \boldsymbol{\mu} \times \mathbf{H}_0 = \gamma\hbar\mathbf{J} \times \mathbf{H}_0 \tag{1-9}$$

The change in **J** during dt, $d\mathbf{J} = \gamma\hbar\mathbf{J} \times \mathbf{H}_0\, dt$, is perpendicular to the plane defined by the vectors **J** and $\mathbf{H}_0$. The motion is a precession, with **J** defining a cone with the axis $\mathbf{H}_0$. The angle between **J** and $\mathbf{H}_0$ remains constant. The precession frequency, $\omega_0 = \gamma H_0$, is known as the Larmor frequency.

In the molecular beam resonance method, H_0 is a spatially homogeneous field produced by the C magnet. In addition, a transverse alternating field $H_1(t)$ is applied at frequency ω in the C magnet region.

$$\mathbf{H}_1(t) = 2\hat{\mathbf{i}}H_1 \cos \omega t \tag{1-10}$$

In the presence of the alternating field, the equation of motion is

$$\frac{d\mathbf{J}}{dt} = \gamma\mathbf{J} \times (\hat{\mathbf{k}}H_0 + \hat{\mathbf{i}}2H_1 \cos \omega t) \tag{1-11}$$

The problem is to find $\mathbf{J}(t)$. Approximate solution to (1-11) can be obtained easily by transforming to an appropriate rotating coordinate system. There is a theorem, proven in all classical mechanics courses, to the effect that the time derivative of $\mathbf{J}$ as viewed from a rotating coordinate system, $\partial\mathbf{J}/\partial t$, is related to $d\mathbf{J}/dt$, the time derivative as viewed from a stationary coordinate system, by the expression

$$\frac{d\mathbf{J}}{dt} = \frac{\partial\mathbf{J}}{\partial t} + \boldsymbol{\omega} \times \mathbf{J} \tag{1-12}$$

where $\boldsymbol{\omega}$ is a vector whose magnitude gives the angular frequency of rotation of the rotating system and whose direction is the axis about which the system rotates. Some necessary insight is obtained by substituting (1-12) into (1-9):

$$\frac{\partial\mathbf{J}}{\partial t} + \boldsymbol{\omega} \times \mathbf{J} = \gamma\mathbf{J} \times \mathbf{H}_0$$

or (1-13)

$$\frac{\partial\mathbf{J}}{\partial t} = \gamma\mathbf{J} \times \left(\mathbf{H}_0 + \frac{\boldsymbol{\omega}}{\gamma}\right)$$

That is, as far as the rate of change of $\mathbf{J}$ is concerned, transforming to a rotating reference system at $\boldsymbol{\omega}$ is the same as adding an effective field $\boldsymbol{\omega}/\gamma$, and considering the motion in the effective field

$$\mathbf{H}_{\text{eff}} = \mathbf{H}_0 + \frac{\boldsymbol{\omega}}{\gamma} \tag{1-14}$$

It follows immediately that $\mathbf{J}$ is time independent in a rotating coordinate system such that $\mathbf{H}_{\text{eff}} = 0$, or $\boldsymbol{\omega} = -\gamma\mathbf{H}_0$. The result, and the transformation, are intuitive for this case. The sense of rotation of $\mathbf{J}$, as seen in the laboratory (i.e., stationary) frame in Fig. 1-4, is just the sense of rotation of the rotating coordinate system necessary to "stop" the motion.

The solution of the problem with alternating (or, conventionally, radio frequency or rf) field, Eq. (1-11), can be obtained from the rotating coordinate transformation after observing the following: decompose $H_1(t)$ into two circularly polarized components of equal amplitude rotating in opposite directions in the xy plane.

$$\mathbf{H}_1(t) = \mathbf{H}_a + \mathbf{H}_b \tag{1-15}$$

where

$$\mathbf{H}_a = H_1(\hat{\mathbf{i}} \cos \omega t + \hat{\mathbf{j}} \sin \omega t)$$

and

$$\mathbf{H}_b = H_1(\hat{\mathbf{i}} \cos \omega t - \hat{\mathbf{j}} \sin \omega t) \tag{1-16}$$

Simple inspection shows that $\mathbf{H}_b$ is rotating in the same sense as the moment in Fig. 1-4, and $\mathbf{H}_a$ in the opposite sense. A rotating coordinate transformation to a system defined by $\boldsymbol{\omega} = \hat{\mathbf{k}}\omega$ leaves the field H_b stationary in the transformed system, but the component rotating in the opposite sense rotates at 2ω in the rotating coordinate system. To see that we may neglect $\mathbf{H}_a$ under most circumstances, examine the effect of $\mathbf{H}_b$ alone. Figure 1-5 shows the fields $\mathbf{H}_0$ and $\mathbf{H}_b$ as they appear in the rotating system

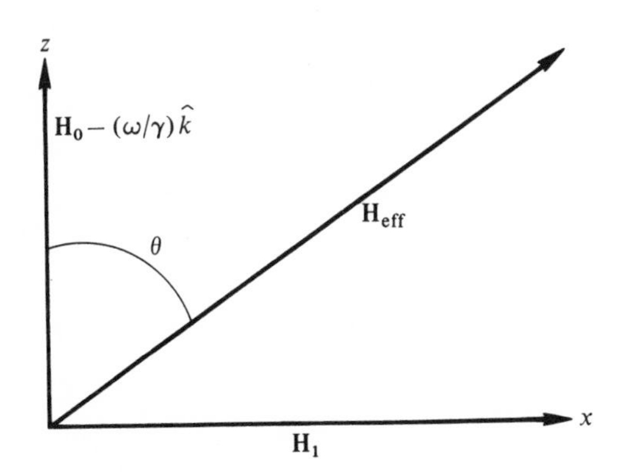

Fig. 1-5 Fields and angles in the coordinate system rotating at $\boldsymbol{\omega} = -\hat{\mathbf{k}}\omega$.

in which $\mathbf{H}_b$ appears stationary, that is, the rotating system defined by the transformation $\boldsymbol{\omega} = -\hat{\mathbf{k}}\omega$.

The effective field in the z direction is $\hat{\mathbf{k}}(H_0 - \omega/\gamma)$ from Eq. (1-14). The total effective field is given by

$$\mathbf{H}_{\text{eff}} = \hat{\mathbf{k}}\left(\mathbf{H}_0 - \frac{\omega}{\gamma}\right) + \hat{\mathbf{i}}\mathbf{H}_1 \tag{1-17}$$

The angle θ is defined by

$$\tan \theta = \frac{H_1}{H_0 - \omega/\gamma} \tag{1-18}$$

The motion of a magnetic moment initially along the z direction consists, in the rotating system, of precession about $\mathbf{H}_{\text{eff}}$ with an angular frequency $\omega_{\text{eff}} = \gamma H_{\text{eff}}$ with the angle θ constant. That is, the vector $\boldsymbol{\mu}$ precesses on the surface of a cone having angle θ and axis $\mathbf{H}_{\text{eff}}$.

A brief digression from the main line of the development is necessary to dispose of the counterrotating component $\mathbf{H}_a$. The fields we have considered produce their effect by virtue of being static in the rotating frame. $\mathbf{H}_a$ rotates at 2ω in that frame, which means that whereas it acts to produce the same general effect as $\mathbf{H}_b$ when it is approximately parallel to it, it produces the opposite effect at a time $\pi/2\omega$ later, when it is directed opposite to $\mathbf{H}_b$. On the average, we expect its effect to be zero. This result has several consequences. First, it is only necessary to apply a linearly polarized rf field in the C region, for we shall be certain, regardless of the algebraic sign of γ, that one of the rotating components will produce some precession of the moment away from the z direction. We arbitrarily chose the magnitude of the linearly polarized field to be $2H_1$, so that the rotating component's amplitude would be the conventional H_1. We could have applied a circularly polarized rf field, and that procedure is the one commonly used to determine the sign of γ.

The arguments we have used to dispose of the counterrotating component were qualitative ones. Quantitatively, the neglect of the counterrotating component is valid only if $(H_1/H_0) \ll 1$. Experiments have been done in fields for which this inequality is not well obeyed. The result is a shift in the radio frequency that produces the maximum effect in tipping the moment to the $-z$ direction. It goes by the name "Bloch–Siegert shift."

For future work, it is important to notice here the obvious fact that as the moment begins to precess around $\mathbf{H}_{\text{eff}}$, it acquires a component in the xy plane. In the laboratory frame, that component is rotating at the angular frequency ω. To make the connection with the magnetic resonance experiment done in the C-field region, between two Stern–Gerlach apparatuses, we must discuss transitions between m_J states, something the preceding discussion of a classical moment seemingly contains no hint of. Let us specialize the discussion to the case of a spin $\frac{1}{2}$ system. The orientation of the classical moment along the z direction is related to the probability that the spin is in the $+\frac{1}{2}$ or $-\frac{1}{2}$ m_J state. If the spin wave function is $|\chi\rangle = a|\frac{1}{2}\rangle + b|-\frac{1}{2}\rangle$, then $P(\frac{1}{2})$, the probability the spin is in the $|\frac{1}{2}\rangle$ state, is $|\langle\frac{1}{2}|\chi\rangle|^2 = |a|^2$ and $P(-\frac{1}{2}) = |\langle-\frac{1}{2}|\chi\rangle|^2 = |b|^2$. We consider the function,

$$\cos\alpha = |a|^2 - |b|^2 \qquad 0 < \alpha < \pi \tag{1-19}$$

with the subsidiary normalizing condition that $|a|^2 + |b|^2 = 1$ (there is certainty of finding the spin in one or the other of the states). The function

$\cos\alpha$ has, in fact, the properties of the projection of the spin on the z axis, and it is through $\cos\alpha$ that we make the connection to the classical calculation. Again, defining $\cos\alpha = \boldsymbol{\mu}\cdot\mathbf{H}_0/|\mu|\,|H_0|$, we find, after some three-dimensional geometry,

$$\cos\alpha = 1 - 2\sin^2\theta\sin^2\frac{\gamma H_{\text{eff}}\,t}{2} \tag{1-20}$$

where we assumed the boundary condition $\alpha = 0$ at $t = 0$. Using the normalizing condition $|a|^2 + |b|^2 = 1$, we can also write (1-19) as

$$\cos\alpha = 1 - 2|b|^2 \tag{1-21}$$

Comparison of (1-20) and (1-21) yields immediately

$$P(-\tfrac{1}{2}) \equiv |b|^2 = \sin^2\theta\sin^2\frac{\gamma H_{\text{eff}}\,t}{2} \tag{1-22}$$

for the probability that, at time t, the spin is in the state $m_J = -\frac{1}{2}$, having started in state $m_J = \frac{1}{2}$ at $t = 0$.

Equation (1-22) checks the result obtained by inspection of Fig. 1-4; namely, that for $\theta = 90°$, the spin precesses to the $-z$ direction in a time $t = \pi/\omega_{\text{eff}} = \pi/\gamma H_{\text{eff}}$. Of course, $\theta = 90°$ corresponds to the condition of exact resonance,

$$\omega = \omega_0 = \gamma H_0 \tag{1-23}$$

It is clear from (1-21) and (1-22) that the probability of a spin being in either of the $m_J = \pm\frac{1}{2}$ states varies periodically, with a full cycle completed with angular frequency γH_{eff}. What is less clear, since we have not done a full quantum mechanical treatment but have only made a plausible connection to quantum mechanics, is that the amplitudes $a(t)$ and $b(t)$ of the $m_J = +\frac{1}{2}$ and $m_J = -\frac{1}{2}$ states, respectively, vary in time in a coherent fashion. The significance of that statement is that the transverse components of angular momentum operators, J_x and J_y, have nonzero but time dependent values. For our future discussion of transient methods in magnetic resonance, as well as the discussion of the Ramsey modification of the molecular beam resonance method, it is most interesting to imagine doing the experiment at exact resonance ($\sin\theta = 1$), and turning off H_1 when $\omega_{\text{eff}}\,t = \pi/2$. Then $P(\frac{1}{2}) = P(-\frac{1}{2}) = \frac{1}{2}$. In the rotating frame, the amplitudes a and b are constant and equal in magnitude (but, since they are complex, not necessarily equal in phase). We can write the spin wave function $|\chi\rangle$ to be $|\chi\rangle = (2)^{-1/2}\,[|\tfrac{1}{2}\rangle + e^{+i\phi}\,|-\tfrac{1}{2}\rangle]$. The expectation value of J_x is, if we suppress the time dependence from the Larmor precession,

$$\begin{aligned}\langle J_x\rangle &= \tfrac{1}{2}[\langle\tfrac{1}{2}| + e^{-i\phi}\,\langle -\tfrac{1}{2}|]J_x[|\tfrac{1}{2}\rangle + e^{i\phi}\,|-\tfrac{1}{2}\rangle] \\ &= \tfrac{1}{2}[\langle\tfrac{1}{2}|J_x|-\tfrac{1}{2}\rangle e^{i\phi} + \langle -\tfrac{1}{2}|J_x|\tfrac{1}{2}\rangle\, e^{-i\phi}] \\ &= \frac{\cos\phi}{2}\end{aligned} \tag{1-24}$$

where ϕ is the phase difference between a and b. It would, in fact, be zero for the example we have used. The important point is that, just as in the classical case, the angular momentum has been turned over toward the xy plane. Viewed in the laboratory frame, it is precessing at ω_0, and will continue to do so indefinitely *with the same phase* as long as nothing perturbs it. The point to emphasize is that the transverse components of $\mathbf{J}$, J_x and J_y, arise from coherent linear superpositions of state functions of the various m_J states. Later, when we deal with ensembles of spins, the question of the relative coherence of the superposition from one spin to the next will be crucial.

The Rabi molecular beam magnetic resonance experiment is done by applying a transverse rf magnetic field in the C magnet. The original, and still common, method of applying the rf field is by means of the "hairpin," shown in Fig. 1-6. The intensity of the beam at the detector is monitored as a function of the radio frequency ω. A typical intensity $I(\omega)$ versus frequency curve has the bell-shaped appearance of Fig. 1-7,

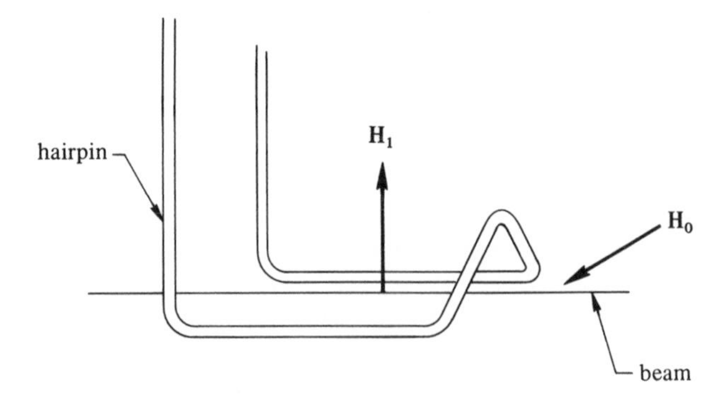

Fig. 1-6 The "hairpin," a method of applying the rf field in the C magnet of a resonance experiment.

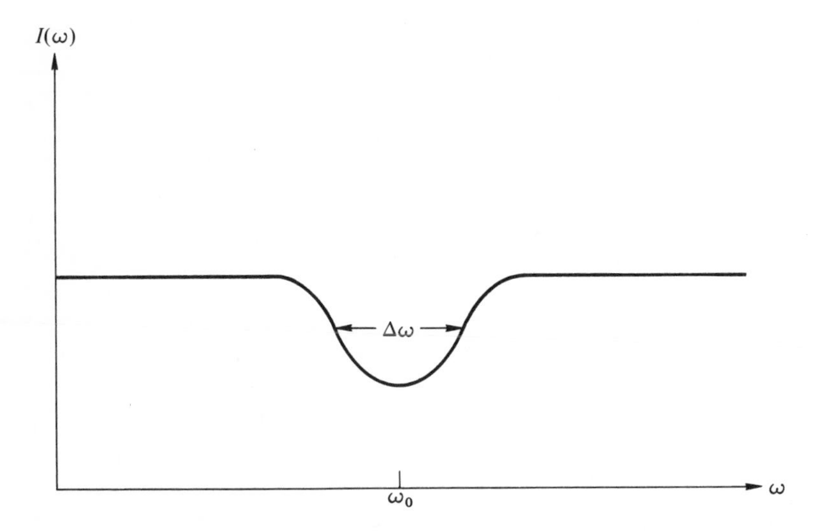

Fig. 1-7 Beam intensity at the detector as a function of the frequency of the rf field in the *C* magnet.

which has been drawn for a hypothetical "flop-out" experiment. The exact shape of the curve is of no particular interest to us, but there is something to be learned by examining the origins of its width, $\Delta\omega$. Since the first object of a beam resonance experiment is to find ω_0, it is important to make $\Delta\omega$ as small as possible. Experimentalists sometimes use the rule of thumb that the frequency ω_0 can be located to within $\Delta\omega$ divided by the signal-to-noise ratio. (Of course, we showed no noise in Fig. 1-7, but it is inevitably there in experimental data.) In the natural effort to make the signal as large as possible, to obtain the optimum "flop-out," the experimenter adjusts the magnitude of H_1 so that, in language appropriate to a $J = \frac{1}{2}$ system, the opposite spin state has the largest possible amplitude at $\omega = \omega_0$. From Eq. (1-22), that occurs when $\gamma H_1 t = \pi$. The time t here is the time of flight of the moment through the hairpin. Approximately half the amplitude of the signal occurs when ω is such that $P(\frac{1}{2}) = P(-\frac{1}{2}) = \frac{1}{2}$. We assume t and H_1 are fixed by the criterion $H_1 t = \pi/\gamma$. Then the half-maximum intensity occurs for $\cos\alpha = 0$ in Eqs. (1-20) or (1-21), or $P(-\frac{1}{2}) = |b|^2 = \frac{1}{2}$ in (1-22). We leave it as an exercise for the reader—an exercise involving primarily the solution of a transcendental equation—to show that $P(\frac{1}{2}) = P(-\frac{1}{2}) = \frac{1}{2}$ when $H_{\text{eff}} = 1.275H_1$, or $\theta = 51.5°$, and $(H_0 - \omega/\gamma) = 0.8H_1$. It follows that the full width at half-maximum intensity, the $\Delta\omega$ of Fig. 1-7, is $\Delta\omega = 1.6\gamma H_1$. When it is coupled with the other requirement, which we have expressed as $\gamma H_1 t = \pi$, we obtain

$$\Delta\omega t = 1.6\pi \tag{1-25}$$

Equation (1-25) looks very much like the uncertainty principle, which relates the precision with which the energy may be established to the time over which the measurement is made. That is exactly what Eq. (1-25) is, for t is the time during which the system is in the probing field H_1. Correct use of the uncertainty principle argument would have obviated the somewhat tedious calculation of the line width, but the effort was worthwhile since it was instructive.

To achieve greater precision, the experimenter must increase t, the time in the apparatus. That goal can be accomplished by selecting the slowest molecules coming from the oven—but with the certain result of a loss in intensity—and also by increasing the length in the beam direction of the C-field magnet and the hairpin. The limitation in $\Delta\omega$, soon reached, is not from the preceding considerations, but rather from the inhomogeneity of the C field. The larger the magnet the harder it is to produce one with a homogeneity of field that is less than the natural limitations of Eq. (1-25). To overcome this limitation, Ramsey devised a very beautiful technique, which we examine briefly for its own sake and for what it can contribute to the understanding of later magnetic resonance experiments.

Consider a monochromatic beam (i.e., constant velocity v) so that each molecule in the beam spends the same time in the hairpin. Ramsey split the hairpin into two parts, both driven by an rf oscillator such that the phase of the rf field is the same in each. They are both in the C magnet, but situated at opposite extremities of the homogeneous field. Figure 1-8

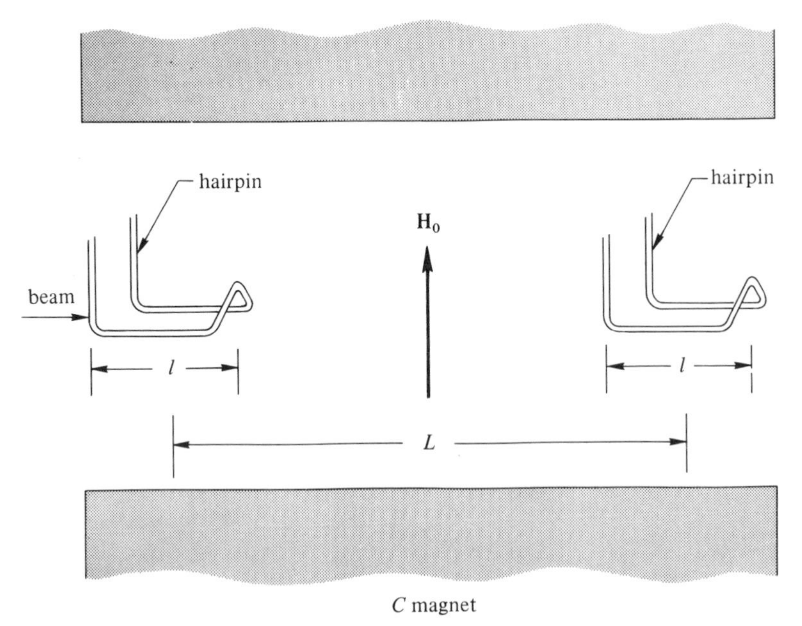

Fig. 1-8 Ramsey split rf field experiment.

shows the arrangement and defines some of the quantities we need. The length L is much greater than l. The latter is adjusted so that $\gamma H_1 t_l = \pi/2$, where $t_l = l/v$. If the beam enters the hairpin in the $m_J = \frac{1}{2}$ state, it leaves in an equal admixture of $m_J = \frac{1}{2}$ and $-\frac{1}{2}$ states, with phase coherence in the admixture. That is, the angular momentum is in the transverse plane and precesses freely about H_0 at $\omega = \gamma H_0$. In the time $t_L = L/v$ it takes for the beam to reach the other hairpin, it precesses an angle Φ in the xy plane. We define the average field $\overline{H}_0$ and the average frequency $\overline{\omega}_0$ by the relation

$$\Phi = \overline{\omega}_0 t_L = \gamma \overline{H}_0 t_L \tag{1-26}$$

The average $\overline{H}_0$ is a spatial average of the fields that the beam sees. (We have assumed for simplicity that the beam has zero transverse dimensions, so that each moment in the beam samples the same H_0.) Adjust the frequency of the oscillator driving the hairpins to be exactly $\overline{\omega}_0$. Then the rf phase of H_1 at the second hairpin is the same as the phase of the transverse component of $\mathbf{J}$; that is, they have the same spatial orientation, so that in the frame rotating at ω_0 the moment continues the precession from the $+z$ to the $-z$ direction that it started in the first hairpin.

Figure 1-9 illustrates the precessions schematically in the rotating reference system. The extent of the improvement of this technique over the single rf field region method can be estimated by the appropriate uncertainty principle argument. The line width $\Delta\omega_{\text{Ramsey}}$ would have to obey

$$\Delta\omega_{\text{Ramsey}}\ t_L \simeq 2\pi \tag{1-27}$$

by analogy with (1-25). Hence, $\Delta\omega_{\text{Ramsey}}/\Delta\omega_{\text{conventional}} = t_l/t_L = l/L$. To see that t_L is indeed the appropriate "measurement time" to insert in an uncertainty principle argument, examine what happens when the radio frequency is not quite ω_0, but is $\omega_0 + \Delta\omega_R$. We now define $\Delta\omega_R$ precisely by noting, as suggested in Fig. 1-9b, that if, during the time t_L, the rf oscillator accumulates phase $\Phi \pm \pi$, then the $\mathbf{H}_1$ seen by the spins in the second hairpin is in the $-y$ direction in the frame rotating at ω_0. The precession of the spins about that field undoes the work of the first hairpin, so the spins precess up to the $+z$ direction again. If precession to the $-z$ direction gives a maximum at the detector, then this second case corresponds to an absolute minimum in the signal. The rf oscillator frequency corresponding to this minimum signal is just

$$(\overline{\omega}_0 + \Delta\omega_R)t_L = \Phi \pm \pi \tag{1-28}$$

or

$$\Delta\omega_R = \pm\pi/t_L$$

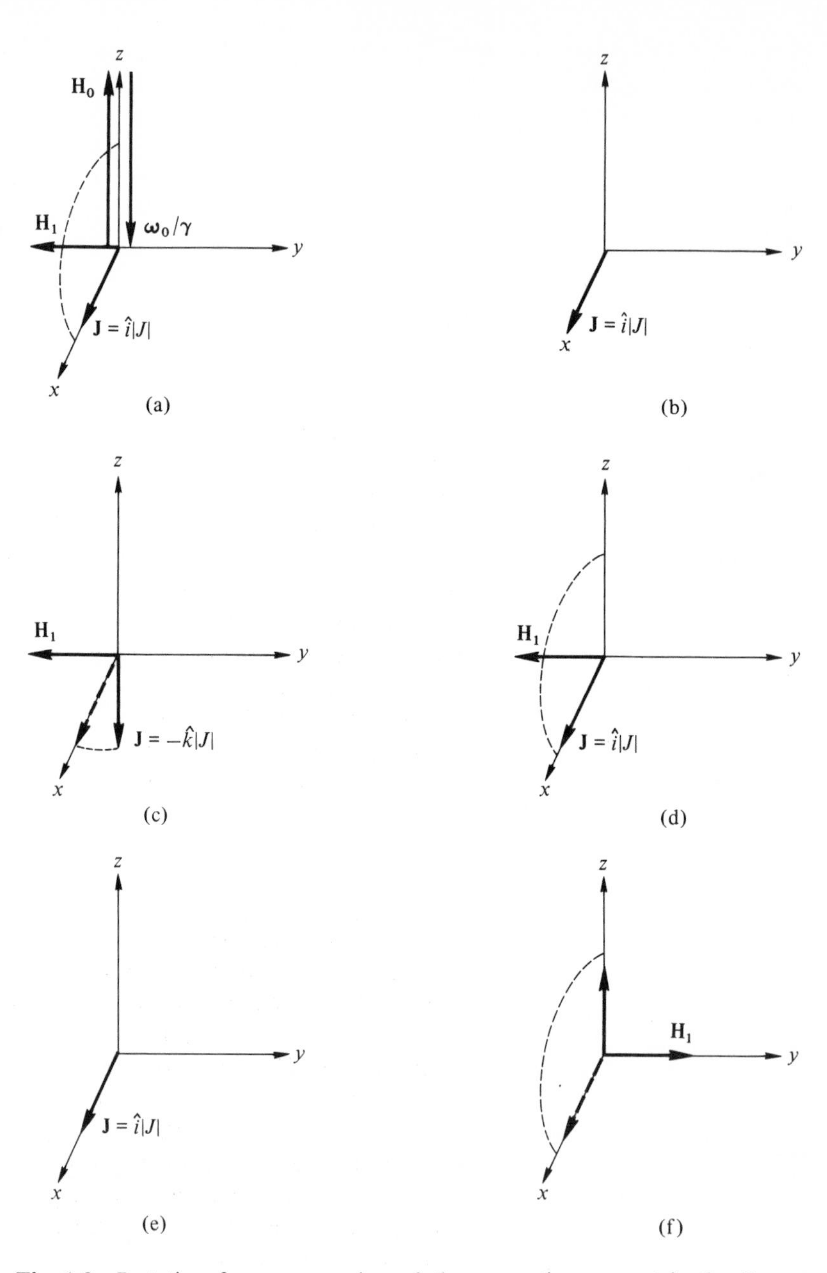

Fig. 1-9 Rotating frame precession of the magnetic moment in the Ramsey split rf field experiment. (a) Precession in the first hairpin. (b) Average orientation of **J** between hairpins. (c) Motion in second hairpin. (d), (e), and (f) Same as (a), (b), and (c) except for reversal of rf phase at second hairpin. See text.

The full width between the minima is $\Delta\omega_R = \pm 2\pi/t_L$. Although $\Delta\omega_R$ calculated this way is almost the same as the "blind" extension of (1-25), not too much should be made of the fact since different quantities are being calculated, which depend in detail on the shape of the resonance line. The shape of a Ramsey curve for a "flop-in" experiment is shown in Fig. 1-10. The width $\Delta\omega$ is roughly the width obtained if the split rf fields are put together.

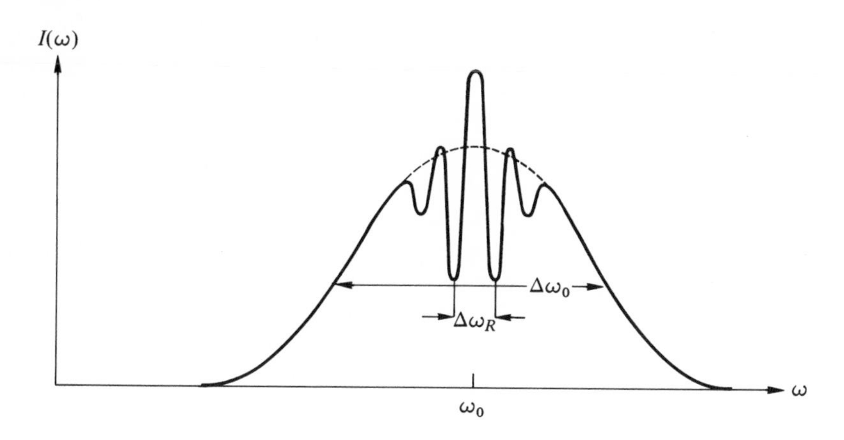

Fig. 1-10 Detector intensity as a function of frequency for the Ramsey experiment.

1-5. APPLICATIONS AND LITERATURE SURVEY

For precision measurements of magnetic moments, the simple Stern–Gerlach apparatus was almost totally eclipsed by resonance methods. The molecular beam techniques by themselves have been used, however, in some important investigations. One of the most obvious is the investigation of the velocity distribution of the beam emitted from the hole of an oven at temperature T. The last publication by Stern before his retirement was an investigation of this subject, which, incidentally, was the topic that prompted his interest in molecular beams in the first place. In the paper by Estermann *et al.* [6], the analysis of the velocities of the molecules is done by measuring their *fall*, or downward deflection, in the earth's *gravitational* field! One rarely encounters any practical consequence of an atomic particle's *gravitational* mass in laboratory atomic physics.

The combined Stern–Gerlach and magnetic resonance experiments of Rabi gave the first precise measurements of nuclear moments, and they are still used for this purpose, particularly, in recent years, to measure nuclear moments of radioactive nuclei. The number of applications is so large as to defy even reasonable enumeration. The student would be advised to read the elementary review articles of Frisch [7] and of Kusch [8], as well as to look into some of the comprehensive tomes, such as Ramsey [1]. Reading the original literature in this field provides a palatable introduction to scientific literature in journal form. Much of it appeared in the 1930's, when brevity to the point of total obscurity was not yet the hallmark of most papers in contemporary journals. The first comprehensive discussion by Rabi *et al.* [9] of the molecular beam, magnetic resonance method, and the first measurement of the anomalous moment of the electron [10] both are relatively readable by upper-division students with vector-model type command of atomic physics. Some further applications of particular importance will be discussed in Chapter 6.

I cannot resist concluding this chapter by remarking on the central position occupied by the Stern–Gerlach experiment in modern quantum physics. The reason seems not to be any particular uniqueness or profundity of the technique, but rather the simplicity of the technique as an example of the problem of state preparation in quantum mechanics. It serves not only as a favorite pedagogical vehicle (see Feynman, vol. 3 [5]) but also as a source of "*gedanken* experiments" for weighty discussion of such vexing questions as the problem of measurement in quantum mechanics. For an example of the latter, see Wigner, *Symmetries and Reflections* [11], particularly p. 160, where the student should experience a shock of recognition if he has read footnote 2. There is no doubt that the Stern–Gerlach experiment, and the Rabi resonance experiment, represent the most elegant, simple, and yet most profound physics experiments of our century.

Problems

1-1. (a) Calculate the vacuum required (in mm Hg) for an atomic beam experiment if the distance from oven to detector is 1 m. Express the result in terms of the cross section σ for collision between a beam atom and a molecule of residual gas. Assume that $\sigma = 10^{-15}$ cm^2 to obtain a quantitative answer.

(b) The atoms in the beam emitted from the oven do not have the same velocity distribution as the atoms in the oven. Show that the beam atoms have a velocity distribution proportional to $v^3 \exp[-3mv^2/2kT]$, where v is the velocity, m is the atomic mass, $k = 1.38 \times 10^{-16}$ ergs°K is Boltzmann's constant, and T is the absolute temperature. Find the most probable velocity and the velocities v_1 and v_2 such that half the atoms in the beam have velocities between v_1 and v_2. Let $T = 500$°K.

(c) Assume the field gradient in the magnet is 10^3 G/cm, and that the magnet is 40 cm long. Assume further that the detector is 50 cm beyond the end of the field gradient region. Find the separation of the two beams of silver atoms for the atoms with the most probable velocity in the beam. Assume $T = 500°K$, and that the width of each beam is as determined in part (b).

1-2. Let the hairpin of Fig. 1-6 be 10 cm long, the separation between the wires 3 mm, and the wires 1 mm in diameter. Find the rf current in the wires necessary to produce a transition from the $m_J = \frac{1}{2}$ to $m_J = -\frac{1}{2}$ state of the ground state of silver for a beam atom of velocity 10^5 cm/sec. What is the width of a resonance line in an apparatus operated under these conditions?

1-3. How far below line of sight does a cesium atom of most probable velocity fall in a 2-m horizontal atomic beam apparatus if the oven temperature is 100°C?

1-4. In our discussion of the Ramsey split field modification of the molecular beam resonance experiment, we assumed the beam molecules to have a single velocity. Discuss qualitatively the consequences to the line shape (Fig. 1-10) if the beam is not monochromatic. Do not forget there are two transit times that may have somewhat different consequences: the time spent in the constant field region between the split rf fields, and the time spent in the rf field region.

References

1. A. Sommerfeld, *Atombau und Spectrallinien*, 4th ed., fr. Vieweg und Sohne, Braunschweig, Germany (1924), p. 145, or 8th ed. (1960), vol. 1, p. 135.
2. N. F. Ramsey, *Molecular Beams*, Oxford University Press, London (1955).
3. H. Kopfermann, *Nuclear Moments*, Academic Press Inc., New York (1958).
4. I. Estermann, O. R. Frisch, and O. Stern, *Nature* **132**, 169 (1933).
5. R. P. Feynman, R. B. Leighton, and M. Sands, *The Feynman Lectures on Physics*, Addison-Wesley, Publishing Co., Reading, Massachusetts (1965), vols. 1, 2, 3.
6. I. Estermann, O. E. Simpson, and O. Stern, *Phys. Rev.* **71**, 238 (1947).
7. O. R. Frisch, *Contemp. Phys.* **1**, 3 (1959).
8. P. Kusch, *Phys. Today* **19**, No. 2, 19 (1966).
9. I. I. Rabi, S. Millman, P. Kusch, and J. R. Zacharias, *Phys. Rev.* **55**, 526 (1939).
10. P. Kusch and H. M. Foley, *Phys. Rev.* **74**, 250 (1948).
11. E. P. Wigner, *Symmetries and Reflections*, Indiana University Press, Bloomington (1967).
12. *Advances in Atomic and Molecular Physics*, D. R. Bates and I. Estermann, Eds., Academic Press Inc., New York, vol. 1 (1965); vol. 2 (1966); vol. 3 (1967); and vol. 4 (1968).

CHAPTER 2

Macroscopic Properties of Nuclear Magnetism

The enormous expansion of the basic ideas of the magnetic resonance technique beyond the molecular beam experiments occurred when it was learned how to do experiments on macroscopic quantities of magnetic moments as found in solids and liquids. There is a considerable difference between flipping an isolated spin from up to down in a molecular beam experiment and doing the analogous thing all at once on 10^{22} spins in ponderable matter. We introduce in this chapter the statistical mechanical steps necessary to describe macroscopic magnetization and its interaction with external electromagnetic fields and with the material in which it is imbedded. The important new idea will be the concept of complex susceptibility, and the practical achievement will be the Bloch equations for the behavior of nuclear magnetism in liquids and some discussion of the apparatus of nuclear magnetic resonance.

2-1. THE EQUILIBRIUM DISTRIBUTION

Chapter 1 has described the basic technique for producing transitions between m_J states of isolated spins. The applications of magnetic resonance in chemistry and solid state physics are based on those methods, but they employ different concepts to produce polarization and to detect the resonance. In this section we shall set forth the basic considerations that govern the relative populations of the different m_J levels when a large number of identical magnetic moments interact with a heat reservoir (usually called a "lattice," even when the moments are in a liquid or a gas). We shall be able to make some very general statements about the way thermal equilibrium is attained, and we shall also derive a few of the macroscopic magnetic properties of the sample, such as its magnetization and magnetic, or Zeeman, energy.

For convenience, we discuss nuclear magnetic moments, which can be nuclei of atoms in a solid, a liquid, or a gas. Much of the discussion will apply also to electron paramagnets, but they have some properties that present complications requiring rather more specialized discussion, for which we refer the reader to Pake [1]. In the beginning, we further idealize the system of interest by assuming we can neglect the interaction of the moments with each other. We also make a more significant approximation that renders irrelevant the "statistics" of the particles—whether Bose–Einstein or Fermi–Dirac—by requiring the density of the system to be low enough to allow the use of Maxwell–Boltzmann statistics. As a result, we specifically exclude from consideration in this chapter conduction electrons in metals or in liquid ^{3}He at low temperatures. Both systems require the Fermi–Dirac distribution function at low temperatures. Otherwise, for nuclei, the density of ordinary solid matter is easily small enough to allow the Boltzmann distribution to be valid.

We start as simply as possible. Consider N spin $\frac{1}{2}$ nuclei in a magnetic field H_0, applied in the traditional z direction. The population of the $m_I = \pm\frac{1}{2}$ states are N_+ and N_-, with $N_+ + N_- = N$. Now the nuclei, whether in solid, liquid, or gas, have translational degrees of freedom—kinetic and potential energy. Call these degrees of freedom the lattice. Let the lattice be homogeneous; characterize it by a single constant temperature T. Our only other assumption about the lattice is that its heat capacity is large compared with the magnetic energy of interaction of the nuclear moments with the magnetic field. For low temperatures and high fields, this assumption is, in practice, rather restrictive, and the theory developed here must be redone to treat that case. Just how low a temperature and how high a field will be the subject of a problem. Figure 2-1 shows the energy-level diagram and helps define relevant quantities.

The Zeeman energy of each spin is

$$E = -\gamma\hbar H_0 m_I \tag{2-1}$$

If $\gamma > 0$, the $m_I = -\frac{1}{2}$ state is higher in energy. To say anything more, we must assume that the spin system (i.e., the totality of N spins, each

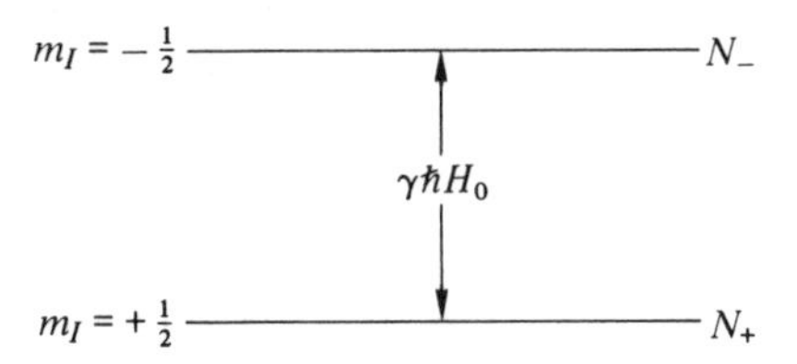

Fig. 2-1 Energy levels of a spin $\frac{1}{2}$ in a magnetic field H_0. Energy levels are ordered for $\gamma > 0$.

interacting with the field H_0) can exchange energy with the lattice. We need not specify the mechanism to make general statements about some of its properties. The lattice must be regarded as a quantum mechanical system with states we label by Greek letters α, β, ... The states are to be thought of as simple harmonic oscillator levels for atoms bound in a solid. The relative occupation of two levels α and β of energy E_α and E_β is proportional to $\exp[(E_\beta - E_\alpha)/kT]$. If the lattice energy consists of the kinetic energy of translation in a gas, for example, then the energies E_α and E_β refer to the kinetic energy, and the preceding expression is Boltzmann's generalization of the Maxwell velocity distribution. The combined system can be labeled in terms of populations of the various states of the subsystems. The populations are specified by (N_+, N_-; N_α, N_β, . . .). Since the combined systems, Zeeman plus lattice, are assumed isolated from the rest of the universe, an increase in Zeeman energy must be accompanied by an equal decrease in lattice energy.

Figure 2-2 shows two lattice state populations that differ by the Zeeman

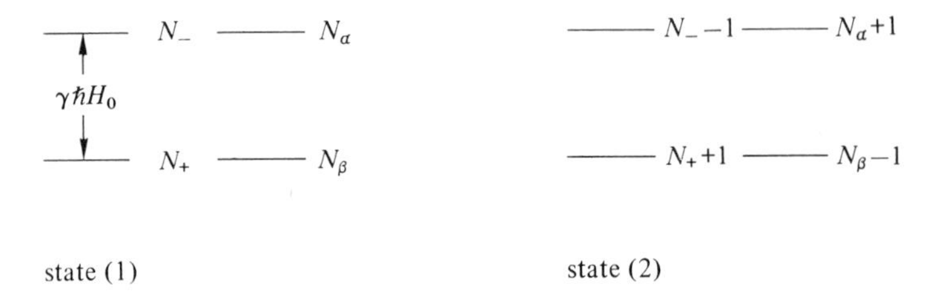

Fig. 2-2 Energy levels for system of spin $\frac{1}{2}$ and a pair of lattice levels with same energy difference. States (1) and (2) of the combined system have the same energy.

energy $\hbar\gamma H_0$, and it indicates schematically two populations of the combined system that have the same energy. Part (1) has the higher Zeeman energy of the two; (2) has the higher lattice energy. We postulate that there is a rate process, determined by a transition rate W, which connects pairs of states the Zeeman populations of which differ by one spin being turned over. We write the following equations for the time rate of change of the spin populations:

$$\frac{dN_+}{dt} = -N_+ W(+ \to -) + N_- W(- \to +) \tag{2-2}$$

and

$$\frac{dN_-}{dt} = -\frac{dN_+}{dt} \tag{2-3}$$

since $N_+ + N_- = N$. Remember that the transition rate $W(+ \rightarrow -)$, for example, involves implicitly a transition from (2) to (1) of Fig. 2-2, *including* the redistribution of lattice-level populations.

The thermal equilibrium condition is $dN_+/dt = 0$, from which we conclude that

$$\frac{N_+^{\,0}}{N_-^{\,0}} = \frac{W(- \rightarrow +)}{W(+ \rightarrow -)} \tag{2-4}$$

where the superscripts indicate thermal equilibrium. On the other hand, the total number of transitions per second of the entire system from (1) to (2) is given by

$$\text{trans/sec from (1) to (2)} = N_- N_\beta w \tag{2-5}$$

where w is a quantum mechanical transition probability involving *only* the squares of matrix elements, densities of states, and constants. The import of the consitution of w is that it is microscopically reversible; w appears in the equivalent expression for the total number of transitions per second from (2) to (1):

$$\text{trans/sec from (2) to (1)} = N_+ N_\alpha w \tag{2-6}$$

In thermal equilibrium, the quantities calculated in (2-5) and (2-6) are equal, from which we conclude

$$\frac{N_+^{\,0}}{N_-^{\,0}} = \frac{N_\beta}{N_\alpha} \tag{2-7}$$

That is, the Zeeman state population ratio is the same as the population ratio of any pair of lattice states separated by the Zeeman energy. This latter ratio is, except for the case of very low temperatures alluded to earlier,

$$\frac{N_\beta}{N_\alpha} = \frac{\exp(-E_\beta/kT)}{\exp(-E_\alpha/kT)} = \exp\frac{-(E_\beta - E_\alpha)}{kT} \tag{2-8}$$

From Eqs. (2-1), (2-7), and (2-8), and from Fig. (2-2), we have

$$\frac{N_+^{\,0}}{N_-^{\,0}} = \exp\frac{-(E_+ - E_-)}{kT} = \exp\frac{+\gamma\hbar H_0}{kT} \tag{2-9}$$

That is, the lower energy level of the spin system, $m_I = +\frac{1}{2}$, is more highly occupied in thermal equilibrium. Moreover, from Eq. (2-3) we get

$$\frac{W(- \to +)}{W(+ \to -)} = \exp \frac{\gamma \hbar H_0}{kT} \tag{2-10}$$

Downward transitions are more probable; indeed, you can look at Eq. (2-10) as providing the mechanism whereby the equilibrium population ratio (2-9) is produced and maintained.

We may now return to (2-3) and talk about the approach to equilibrium from a nonequilibrium initial condition. Define the population difference

$$n = N_+ - N_- \tag{2-11}$$

and rewrite N_+ and N_- in terms of n and N:

$$N_+ = \tfrac{1}{2}(N + n) \tag{2-12a}$$

$$N_- = \tfrac{1}{2}(N - n) \tag{2-12b}$$

In terms of N and n, Eq. (2-3) becomes

$$\frac{dn}{dt} = N[W(- \to +) - W(+ \to -)] - n[W(+ \to -) + W(- \to +)] \tag{2-13}$$

or, by factoring out $[W(+ \to -) + W(- \to +)]$,

$$\frac{dn}{dt} = \frac{n_0 - n}{T_1} \tag{2-14}$$

where

$$n_0 = N \frac{W(- \to +) - W(+ \to -)}{W(+ \to -) + W(- \to +)} \tag{2-15}$$

and

$$\frac{1}{T_1} = W(+ \to -) + W(- \to +) \tag{2-16}$$

where n_0 is the equilibrium population difference, as substitution of (2-10) and (2-11) into (2-15) will verify:

$$n_0 = N \frac{W(+ \to -)[\exp(\gamma\hbar H_0/kT) - 1]}{W(+ \to -)[\exp(\gamma\hbar H_0/kT) + 1]} = N \tanh\left(\frac{\gamma\hbar H_0}{2kT}\right) \quad (2\text{-}17)$$

The time T_1, defined by (2-16), is the spin-lattice relaxation time; it is the time constant of the approach of the spin system to thermal equilibrium with the lattice. If (2-14) is solved with $n(0) = 0$ as the initial condition, as would be the case if the field were switched on suddenly at $t = 0$ after having always been zero before, we find that

$$n = n_0[1 - \exp(-t/T_1)]$$

It is appropriate at this point to put in perspective what we have been doing, and perhaps even allow a glimpse of a skeleton in a closet. Equations (2-3), (2-5), and (2-6) are examples of the *principle of detailed balance*, which was first used by Einstein in his 1916 rederivation of the Planck radiation formula. Given Eq. (2-2), which is also known as a *master equation*, the irreversibility of the approach to equilibrium is already determined, even though (2-2) makes explicit use of the microscopically reversible quantum mechanical transition probability w, introduced in Eq. (2-5). The justification, or, if you wish, derivation of the master equation is the central problem of nonequilibrium statistical mechanics. Magnetic resonance experiments on nuclear spin systems in solids and liquids have in recent years provided interesting and tractable model systems for which specific derivations of equations such as Eq. (2-3) could be tested, and the limitations understood.

Although it is rather far afield from our main purpose, something more than the preceding mysterious remarks can be made with regard to the origin of the irreversibility. In terms of the coefficients a and b introduced in Chapter 1 for the spin wave functions $|\chi\rangle = a|\frac{1}{2}\rangle + b|-\frac{1}{2}\rangle$, it is clear that (2-2) is an equation in $|a|^2$ and $|b|^2$, since the probability of occupation of a given stage for a single spin is essentially $(1/N)$ times the occupation of that state in the ensemble of N identical spins. The information that is missing is the relative phase of the $|\frac{1}{2}\rangle$ and $|-\frac{1}{2}\rangle$ states, which, if we remember Chapter 1, is related to the transverse component of the magnetic moment. In equilibrium there is no transverse macroscopic magnetic moment, not even a coherent alternating one. From that, we reason backward to the conclusion that the relative phase of the $m_I = \pm\frac{1}{2}$ states from spin to spin must be a random quantity, so that the total transverse moment vanishes. The reasoning is purposely circular, but it points to the crux of the problem and reintroduces the idea that a macroscopic transverse magnetization, such as produced in a magnetic resonance experiment, has to do with a coherent admixture, from spin to spin, of the magnetic states m_I.

2-2. ENERGY, MAGNETIZATION, AND SUSCEPTIBILITY

The magnetic energy, or Zeeman energy, of the spin system is given for a general spin I by

$$E = \sum_{m_I=-I}^{I} E(m_I)N(m_I) \tag{2-18}$$

It is convenient to define the zero of energy $E(m_I)$ for each spin to be at $m_I = 0$ for I even, and midway between the $m_I = \pm\frac{1}{2}$ energy levels for I an odd half-integer. Then Eq. (2-1) is the appropriate expression for $E(m_I)$ in Eq. (2-18). For $N(m_I)$, we shall simplify matters a little by using an expression that will be valid in the high temperature limit only: $\gamma\hbar H_0 \ll kT$.

$$N(m_I) = \frac{N\exp(-\hbar\gamma m_I H_0/kT)}{\sum_{-I}^{I}\exp(-\hbar\gamma m_I H_0/kT)} \cong \frac{N}{2I+1}\exp\frac{-\hbar\gamma m_I H_0}{kT} \tag{2-19}$$

The first equation in expression (2-19) is, of course, exact. The denominator is that fundamental expression of statistical mechanics, the "sum over states," or partition function. The replacement of the denominator by $2I + 1$ in the second equation in (2-19), although retaining the full exponential expression in the numerator, is conventional but inconsistent. If $\gamma\hbar H_0 m_I/kT \ll 1$, the exponential in the denominator may be expanded:

$$\sum_{-I}^{I}\exp\frac{-\gamma\hbar H_0 m_I}{kT} = (2I+1) - \frac{\gamma\hbar H_0}{kT}\sum_{-I}^{I} m_I + \frac{1}{2}\left(\frac{\gamma\hbar H_0}{kT}\right)^2 \sum_{-I}^{I} m_I^2 + \cdots$$

The second term in the sum is zero since $\sum m_I = 0$, but the next term is not. Thus, the denominator's lowest term in the expansion parameter is quadratic, but the numerator's lowest term is linear. To be consistent, we must take two terms of the numerator and one of the denominator. The exponent in the numerator has been retained at this stage because one so often is concerned with population ratios, in which case it is somehow easier always to write

$$\frac{\exp(E_\alpha/kT)}{\exp(E_\beta/kT)} = \exp\frac{(E_\alpha - E_\beta)}{kT} \cong 1 + \frac{E_\alpha - E_\beta}{kT}$$

than

$$\frac{1 + (E_\alpha/kT)}{1 + (E_\beta/kT)} \simeq \left(1 + \frac{E_\alpha}{kT}\right)\left(1 - \frac{E_\beta}{kT}\right) \simeq 1 + \frac{E_\alpha - E_\beta}{kT}$$

Equation (2-19) satisfies Eq. (2-9), agrees with (2-17) in the high temperature limit, and satisfies conservation of spins, $\sum_{-I}^{I} N(m_I) = N$, in that limit also. Expansion of (2-18) with (2-19) in the high temperature limit yields

$$E \simeq \frac{N}{2I+1} \sum_{-I}^{I} \gamma\hbar H_0 m_I \left(1 - \frac{\gamma\hbar H_0}{kT} m_I\right)$$

$$= -\frac{N}{2I+1} \left(\frac{\hbar H_0}{kT}\right)^2 \sum_{-I}^{I} m_I^2$$

It is easily verified that $\sum_{-I}^{I} m_I^2 = [I(I+1)(2I+1)]/3$, so that

$$E = \frac{N\gamma^2\hbar^2 I(I+1)H_0^2}{3kT} \tag{2-20}$$

As an aside, it is interesting to compare the formula (2-20) with the classical formula in the same limit:

$$E = -\langle \boldsymbol{\mu} \cdot \mathbf{H}_0 \rangle N = -\mu H_0 N \langle \cos\theta \rangle \tag{2-21}$$

where $\langle\ \rangle$ indicates the thermal average of the quantity inside it, calculated according to Boltzmann statistics. To find $\langle \cos\theta \rangle$, one must weight $\cos\theta$ by the probability that the moment μ is oriented at angle θ to H_0:

$$\langle \cos\theta \rangle = \frac{\int_\Omega \cos\theta \exp(\mu H_0 \cos\theta / kT)\, d\Omega}{4\pi}$$

where the integration is over the solid angle the complete range of which is the denominator 4π. Expand the exponential. The first term vanishes and the second yields:

$$\frac{\mu H_0}{kT} \frac{1}{2} \int_0^\pi d\theta \cos^2\theta \sin\theta = \frac{1}{3} \frac{\mu H_0}{kT}$$

Substituting back into (2-21), we obtain

$$E = \frac{-N\mu^2 H_0^2}{3kT}$$

The quantum mechanical equivalent of μ^2 is $\gamma^2\hbar^2 I(I+1)$,[1] and the factor

[1] Occasionally one sees the quantity $[g^2 I(I+1)]^{1/2}$ defined to be the magnetic moment, rather than gI. The former is most frequently found in the older literature on magnetism.

of 3 in the denominator of (2-20) is the quantum average of m_I^2, just as it is the average of $\cos^2 \theta$ in the classical calculation. (This equivalence was known as the "principle of spectroscopic stability" in the early days of quantum mechanics.)

The magnetic moment per unit volume may be defined by the expression $E/V = -\mathbf{M}_0 \cdot \mathbf{H}_0$. From (2-20), we see that

$$\mathbf{M}_0 = \frac{N}{V} \frac{\gamma^2 \hbar^2 I(I+1)}{3kT} \mathbf{H}_0 \tag{2-22}$$

The same expression is obtained from $M_0 = n_0 \mu_z / V$, where n_0 is given by the high temperature expansion of (2-17). The static magnetic susceptibility χ is defined by $\mathbf{M}_0 = \chi_0 \mathbf{H}_0$, so we obtain the well-known formula

$$\chi_0 = \left(\frac{N}{V}\right) \frac{\gamma^2 \hbar^2 I(I+1)}{3kT} \tag{2-23}$$

A word about units is in order. In the Gaussian system, which is still largely used in research physics even though it is no longer the system of choice in elementary courses, the dimensions of M, B, and H are the same. Hence, χ_0, from the defining expression, is dimensionless. As defined here, it is often referred to as the *volume susceptibility*, however, to distinguish it from *mass* or *molar* susceptibilities. These quantities arise from a definition of M_0 in which the number of moments contributing is not the number in cubic centimeters, as in (2-22), but the number in a gram or the number in a mole. With such a definition, since M_0 and H_0 must still have the same dimensions, the quantity χ_0 is not necessarily dimensionless. The mass and molar susceptibilities are related to the volume (i.e., dimensionless) susceptibility by

$$(\chi)_{\text{mass}} = \frac{(\chi)_{\text{vol}}}{\rho} \tag{2-24a}$$

$$(\chi)_{\text{molar}} = (\chi)_{\text{vol}}\, v \tag{2-24b}$$

where ρ in (2-24a) is the density, and v in (2-24b) is the molar volume, $v = N_A M/\rho$, where N_A is Avogadro's number, and M is the atomic mass.

2-3. RESPONSE TO AN ALTERNATING FIELD; COMPLEX SUSCEPTIBILITIES

In the Rabi magnetic resonance experiment, isolated spins interacted only with the static and alternating fields. Earlier in this chapter we introduced the concept of spin-lattice interaction. We now need to combine them all and discuss, as we did before, the components of the magnetic moment in the presence of these interactions. Now, however, the magnetic moment to consider is the macroscopic magnetic moment of the entire sample. We shall begin by discussing the z component, followed by the transverse components M_x and M_y.

There is a subtle difference between the effect of the rf field on the isolated atom in Eq. (1-22) and its effect in the presence of spin-lattice interaction. The origin of the distinction lies in the fact that with spin-lattice interaction the states m_I are not quite eigenstates of the total Hamiltonian. When one speaks of the states in the approximate language of m_I, one must then absorb the approximate nature of the description into a lack of sharpness of the energy level, and one speaks of the level as having a breadth given, in fact, by the uncertainty principle argument $\Delta E > \hbar/T_1$. In discussing M in the presence of the rf field, we are particularly interested in small deviations of M from M_0. The rf field is regarded as a perturbation that, in the language of Chapter 1, is to produce only a small amplitude b in a state which initially had $|a|^2 = 1$. Thus, in Eq. (1-22), $P(\frac{1}{2}) \ll 1$ in the perturbation theory limit, if $|a|^2 \simeq 1$ is to continue to be true. $P(-\frac{1}{2})$ is small for small times t after the perturbation is turned on, and one sees immediately that $P(-\frac{1}{2}) \simeq t^2$ for intervals of t such that $P(-\frac{1}{2}) \ll 1$. Fortunately for the preservation of a linear theory, the quadratic dependence on t is not correct for the problem with which we are now concerned, and the correct time dependence for the probability of the $m_I = -\frac{1}{2}$ state being occupied is a linear rather than a quadratic one. The apparent paradox is resolved by noting that $P(-\frac{1}{2})$ was computed for exact eigenstates $m_I = \pm\frac{1}{2}$, whereas these are not exact eigenstates in the presence of spin-lattice relaxation. The final result must involve an integration over all the states making up the "level" of nonzero width. The details are carried out in Abragam's treatise *Nuclear Magnetism* [2] (see Chapter 2).

The probability of a transition from the state m_I to the state $m_I - 1$ is proportional to time. Unlike the case of spin-lattice relaxation, however, the transition rate for a spin in state $m_I - 1$ to m_I is the same as the transition rate for a spin in state m_I to $m_I - 1$. The processes are to be contrasted: the spin-lattice relaxation process drives the population difference n toward the thermal equilibrium value n_0; the rf field drives n toward zero. The latter statement is easily verified:

$$\left(\frac{dN_+}{dt}\right)_{\text{rf}} = -N_+ W_{\text{rf}} + N_- W_{\text{rf}} = -nW_{\text{rf}}$$

and

$$\left(\frac{dN_-}{dt}\right)_{\text{rf}} = -\left(\frac{dN_+}{dt}\right)_{\text{rf}}$$

Hence,

$$\left(\frac{dn}{dt}\right)_{\text{rf}} = -2W_{\text{rf}}\, n \tag{2-25}$$

Equilibrium occurs when $n = 0$, Q.E.D. The transition probability per unit time W_{rf} is properly computed by the standard time dependent perturbation formula (the so-called "golden rule") of quantum mechanics. The perturbation is the interaction of the magnetic moment and the rf field, $-\boldsymbol{\mu} \cdot \mathbf{H}_1$; we need remark here only that W_{rf} is proportional to $H_1{}^2$.

The total rate of change of n is the sum of (2-14) and (2-25)

$$\frac{dn}{dt} = -2W_{\text{rf}}\, n + \frac{n_0 - n}{T_1} \tag{2-26}$$

In the steady state $dn/dt = 0$, and (2-26) shows that

$$n = \frac{n_0}{1 + 2W_{\text{rf}}\, T_1} \tag{2-27}$$

Equations (2-26) and (2-27) may be rewritten in terms of $M_z = n\gamma\hbar/2$ simply by substituting M_z for n and M_0 for n_0. Equation (2-27) shows that the rf field does not appreciably disturb M_z from M_0 as long as $2W_{\text{rf}}\, T_1 \ll 1$. When this inequality is not satisfied, $n < n_0$, and the resonance is said to be *saturated.* Additionally, the rate of energy absorption from the rf field may be calculated from (2-26),

$$\frac{dE}{dt} = \hbar\omega n W_{\text{rf}} = \frac{n_0\, \hbar\omega W_{\text{rf}}}{1 + 2W_{\text{rf}}\, T_1} \tag{2-28}$$

Note that dE/dt becomes independent of W_{rf} as W_{rf} exceeds $\frac{1}{2}T_1$.[2]

The discussion of the transverse components of M in the steady state situation must be phrased in rather general terms. It is again convenient to work in the frame rotating with the rf field, angular frequency ω. Let H_1 be along the x axis in the rotating frame. Define susceptibilities χ' and χ'' by the relations

$$M_x = 2\chi' H_1 \tag{2-29a}$$

$$M_y = 2\chi'' H_1 \tag{2-29b}$$

Equations (2-29) are, first of all, linear. That is, the transverse magnetization is proportional to the first power of the perturbing rf field H_1. Note that χ' has been defined to be the proportionality factor between H_1 and

[2] The failure of the first attempts to observe nuclear magnetic resonance in solids, by C. J. Gorter in the late 1930's, can be traced to the use of a sample with a T_1 that probably was several hours, so that $2W_{\text{rf}}T_1$ was undoubtedly very large and the power absorbed from the rf field by the sample was very small.

the component of M parallel to, or in phase with, H_1 in the rotating frame. The out-of-phase or orthogonal component is determined by χ''.

Necessary additional insight is obtained by transforming back to the laboratory frame, X, Y, $Z = z$, where we assume that the rotating H_1 is produced by a linearly polarized rf field in the X direction:

$$H_x(t) = 2H_1 \cos \omega t$$

In the laboratory frame, then, the X component of the magnetization is, from Eq. (2-29),

$$M_X(t) = (\chi' \cos \omega t + \chi'' \sin \omega t)2H_1 \tag{2-30}$$

Equation (2-30) may be expressed more compactly, and more conventionally, as the real part of a complex quantity. We define the complex quantities $\mathscr{H}_X = 2H_1 e^{i\omega t}$ and $\mathscr{M}_X = 2\chi H_1 e^{i\omega t}$. Their real parts are the physical field and magnetization, respectively. Comparison with Eq. (2-30) shows that $M_X = \text{Re}\, \mathscr{M}_X$ only if χ is the complex quantity

$$\chi = \chi' - i\chi'' \tag{2-31}$$

The linear relation between the complex driving term $\mathscr{H}_X(t)$ and the complex response function $\mathscr{M}_X(t)$ has many familiar parallels in physics. Probably the first one encountered by most students is the generalized Ohm's law from ac circuit theory, $\mathscr{V} = \mathscr{I}\mathscr{Z}$, where $\mathscr{Z}$ is the complex impedance. The same care must be used in calculating quantities with complex $\mathscr{M}$ and $\mathscr{H}$, as with complex current, impedance, and voltage. Thus the instantaneous power absorbed from a generator is (Re $\mathscr{M}$) (Re $\mathscr{H}$), not Re($\mathscr{M}\mathscr{H}$).

The average power absorbed by a unit volume of material is

$$P = \frac{1}{T}\int_0^T \text{Re}\, \mathscr{H} \cdot \text{Re}\left(\frac{d\mathscr{M}}{dt}\right) dt \tag{2-32}$$

where $T = 2\pi/\omega$ is an rf period. The only term in the scalar product in the integrand is the X component, so

$$P = \frac{1}{T}\int_0^T (2H_1)^2 \cos \omega t(-\omega\chi' \sin \omega t + \omega\chi'' \cos \omega t)\, dt$$

$$= 4H_1{}^2\omega\chi''\left(\frac{1}{T}\int_0^T \cos^2 \omega t\, dt\right) = 2H_1{}^2\omega\chi'' \qquad (2\text{-}33)^3$$

[3] Comparison with Eq. (2-28) would allow immediate determination of χ'' if W_{rf} were known. We shall not pursue that course, since we shall eventually get W_{rf}, for the particular situation (2-28) represents (lifetime broadened levels) without using quantum mechanical perturbation theory. See the discussion at the end of Section 2-5.

The analogy between the impedance and χ cannot be made blindly. The real part of $\mathscr{Z}$ determines the loss, and the imaginary part determines the nature of the periodic but lossless exchange of energy between the circuit and the generator. The terms "real" and "imaginary" must be interchanged in the preceding discussion because the voltages that interact with currents in the magnetic system are *induced*; they are determined by $d\mathscr{M}/dt = i\omega\mathscr{M}$. The phenomenon of induction accounts for the factor of ω in (2-33) and the appearance of χ'' instead of χ' in the expression for the absorbed power.

Without further assumptions about the details of the system, except for the all-important one that the "cause must precede the effect," one can establish that χ' and χ'' are not independent of each other but are related by integral relations known as the Kramers–Kronig relations, which were independently derived with reference to optical absorption by Kramers and Kronig in 1926. A thorough discussion and derivation of these relations may be found in the text by Slichter [3], in which there is a precise mathematical formulation of the somewhat enigmatic statement made previously about cause and effect. Relations such as those of Kramers and Kronig exist between the real and imaginary parts of the complex linear response functions of physical systems, whether they be magnetic or electric susceptibilities, impedances of passive electrical circuits, or reactions in elementary particle physics.

2-4. THE BLOCH EQUATIONS

Nuclear magnetic resonance in condensed material (namely, hydrogenous materials such as water or paraffin wax) was first observed in 1946 independently by Professor Felix Bloch and coworkers at Stanford, and Professor E. M. Purcell and coworkers at Harvard.[4] In conjunction with the experiments at Stanford, Bloch proposed phenomenological equations of motion for the macroscopic magnetization vector **M**, which serve very well to describe magnetic resonance experiments in liquids, gases, or "liquidlike" solids. It will be one of the tasks of this book to enable the student to comprehend in physical terms the limitations of the Bloch equations, but first we must set them down and explore their solutions.

The object is to express the interaction of the magnetization with the external fields (static and alternating), with the lattice, and to write a term that expresses the interaction of the magnetic moments with each other and with other *internal* magnetic fields in the sample. We have already accomplished the first two tasks, and we have a major portion of the Bloch

[4] Bloch and Purcell shared the 1952 Nobel prize for their work, the third Nobel prize awarded for work described in this book. Stern won the prize for his molecular beam work in 1943, and Rabi for the magnetic resonance method in 1944.

equations if we assemble Eqs. (1-11) and (2-26) in slightly altered and compatible form. The effect of the interaction of the moments undergoing resonance with each other and other magnetic moments in the sample, via the dipole-dipole interaction or through more esoteric quantum mechanical effects called exchange interactions, is contained within the Bloch equations by a single parameter that affects only the transverse magnetization, the components of which are M_x and M_y. (We shall use lowercase subscripts for the laboratory coordinate system in this section, and identify equations written in the rotating system explicitly.)

Bloch assumed that the internal interactions of spins with each other could be expressed by the equation

$$\frac{\partial M_{x,y}}{\partial t} = -\frac{M_{x,y}}{T_2} \tag{2-34}$$

Equation (2-34) defines the parameter T_2, known variously as the transverse or spin-spin relaxation time. (The partial derivative has been written only to call particular attention to the existence of the other terms that cause $M_{x,y}$ to change.) Equation (2-34) is the equation for a magnetization that decays exponentially to zero. Suppose that in the rotating frame at $\omega_0 = \gamma H_0$ the transverse component M_x is created at $t = 0$. In the context of Chapter 1 it persisted indefinitely. What can cause it to decay? One obvious cause would be an inhomogeneous magnetic field across the sample so distributed that the Larmor frequencies of the spins in the various parts of the sample differ sufficiently so that in time T_2 they would get out of phase with each other enough to diminish the initial M_x to $1/e$ of its value. Although it would take special field inhomogeneity to make the magnetization decay exactly exponentially, the point is worth illustrating with a figure. Figure 2-3 shows $M_x(0)$, in the rotating frame. Imagine

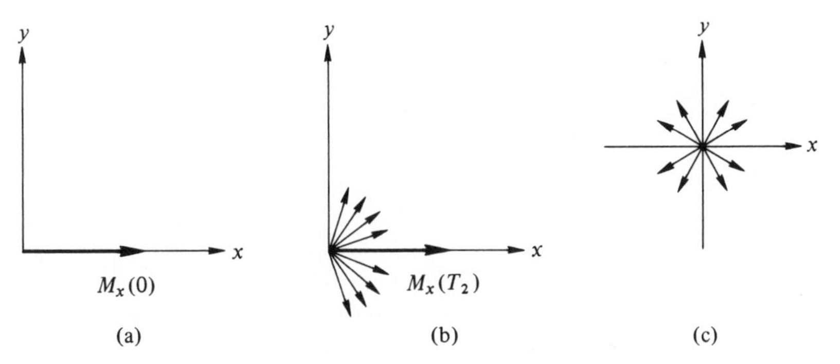

Fig. 2-3 Rotating frame view of decay of M_x. (a) Initial condition. (b) Partial decay ($t \sim T_2$). (c) Total decay ($t \gg T_2$).

M_x made up of small magnetization vectors that precess at a variety of frequencies differing by small amounts either way from the average frequency ω_0. Then in a frame rotating at ω_0, they precess one way or the other until by $t = T_2$, their vector sum is still in the x direction but has diminished to $M_x(0)/e$.

One internal physical process that diminishes the transverse magnetization is the spin-lattice relaxation time T_1. Any other process, such as field inhomogeneity, only adds to the rate at which $M_{x,y}$ diminishes, so that $T_2 < T_1$. The most interesting T_2 process is the interaction of each spin with internal fields. By analogy with the external field inhomogeneity mechanism, we should guess that the internal fields that act to dephase $M_{x,y}$ are those parts of the total internal fields which are in the z direction and which are static and quasistatic. That the internal fields should manifest themselves just as single parameter T_2 in an equation of the form of Eq. (2-34) is, in fact, a result of rather special circumstances which we shall explore later.

Combining Eq. (2-34) with the torque and relaxation equations, we get the Bloch equations:

$$\frac{dM_z}{dt} = \frac{M_0 - M_z}{T_1} + \gamma(\mathbf{M} \times \mathbf{H})_z \tag{2-35a}$$

$$\frac{M_{x,y}}{dt} = -\frac{M_{x,y}}{T_2} + \gamma(\mathbf{M} \times \mathbf{H})_{x,y} \tag{2-35b}$$

where

$$\mathbf{H} = \hat{\mathbf{k}}H_0 + \hat{\mathbf{i}}H_1(t)$$

2-5. SOLUTIONS OF THE BLOCH EQUATIONS

Solutions of Eqs. (2-35) are not difficult to obtain for a few special experimental conditions that are also of particular interest in practice. We have actually already discussed, in two separate parts, one particularly interesting solution that we can do without mathematics; thus we begin with that example.

Free induction decay

In making the plausibility argument for the form of the T_2 term, we began arbitrarily with the magnetization in the x direction in the rotating frame. The subsequent exponential decay with time constant T_2, determined by Eq. (2-34), appears in the laboratory frame as

$$M_x(t) = M_{x0} \cos \omega_0 t \exp \frac{-t}{T_2} \tag{2-36}$$

We have two problems: how to produce $M_x(0)$ and how to detect $M_x(t)$. Producing $M_x(0)$ may be accomplished by a "90° pulse," a transverse rf pulse of magnitude H_1 in the rotating frame at $\omega_0 = \gamma H_0$, which acts for a time τ such that $\gamma H_1 \tau = \pi/2$. From Chapter 1 we see that if H_1 is stationary in the y direction in the rotating frame, $M_z = M_0$ precesses about H_1 at $\gamma H_1 = \omega_1$ until, at $\tau = \pi/2\gamma H_1$, it is pointing in the $-x$ direction. The experimental arrangement to do this is shown in Fig. 2-4. We have neglected one thing of great importance. The pulsed rf field that produces the 90° rotation of M_0 in the rotating frame must act in the presence of T_1 and T_2 processes, rather than in their absence, as in Chapter 1. Consequently, the 90° nutation of M_0 is only an approximate description of what happens, and we must find how good an approximation it is. If the major torque on the magnetization is to be from H_1 during $0 < t < \tau$, then the relaxation toward H_0 (in time T_1) and the dephasing of the transverse magnetization (in time T_2) must not be important compared to the precession about H_1 during τ: $\tau \ll T_2 < T_1$. The requirement on H_1 is thus $(\pi/2\gamma H_1) \ll T_2$, or

$$\gamma H_1 \gg \frac{\pi}{2T_2} \tag{2-37}$$

It is the same result we get if we assume, plausibly, that H_1 must be much larger than the internal fields or external field inhomogeneities described by T_2.

A word is in order about the magnitudes involved for a nuclear resonance experiment. The largest internal fields in ordinary substances (think of NaCl, for example) are caused by the nuclear magnetic dipole-dipole interaction. The magnetic field produced by one dipole a distance r from another is on the order of μ/r^3. Typically, since $\mu \cong \gamma\hbar = 10^{-23}$ ergs/G, and $r \sim 2 \times 10^{-8}$ cm, $\Delta H = \mu/r^3 \cong 10^{-23}/8 \times 10^{-24} \cong 1$ G. Although the Bloch equations are not, in fact, generally valid for solids, they do provide a framework for rapid estimates of upper or lower limits. To satisfy our inequalities, $H_1 \gtrsim 10$ G is required, and $\tau < \pi/(2 \times 10^4 \times 10) \cong 10$ μsec. The T_2 for this case is about $T_2 \sim 1/\gamma\,\Delta H = 100$ μsec. These calculations provide upper limits on fields and lower limits on times for this particular substance, because, as we shall see in the next chapter, the effect of the nuclear motion that occurs in a liquid or gas is to decrease the effective dipole-dipole interaction for T_2 processes, often by several orders of magnitude.

How large is the induced signal? The precessing magnetization in the xy plane in the laboratory frame is of initial magnitude $M_0 = \chi_0 H_0$, where χ_0 is the static susceptibility, Eq. (2-23). Referring to Fig. 2-4, let the coil that produced the 90° pulse of H_1 serve also to pick up the induced

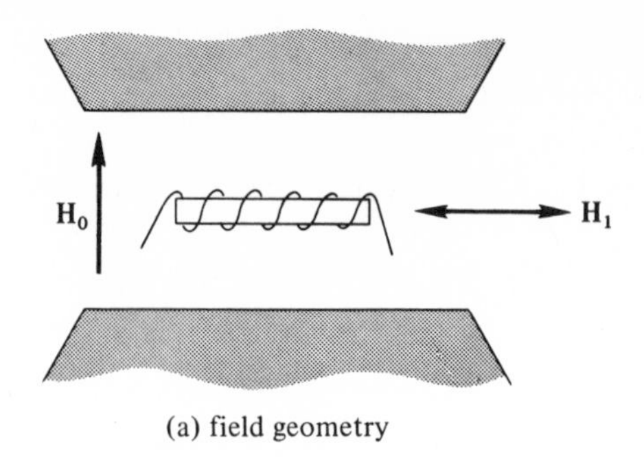

(a) field geometry

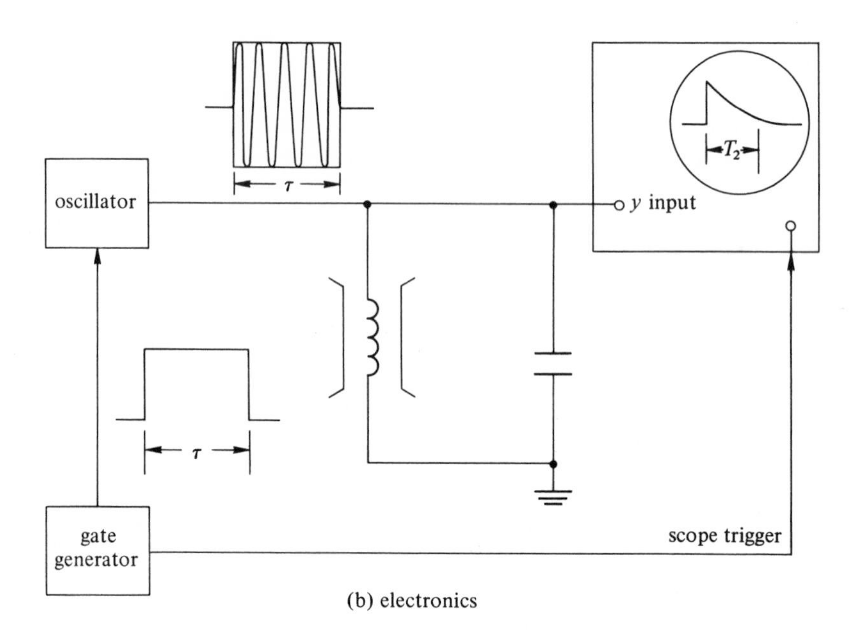

(b) electronics

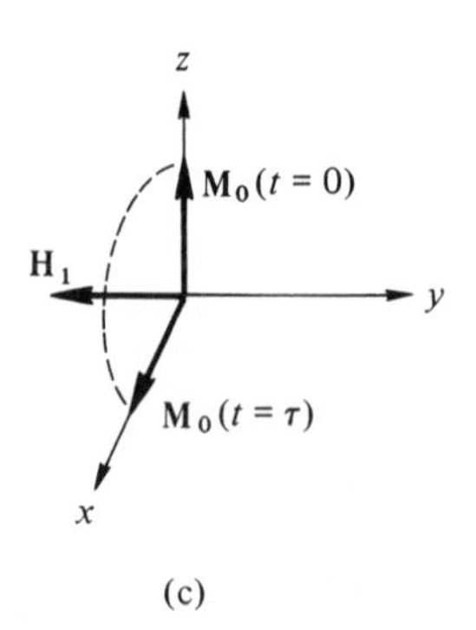

(c)

Fig. 2-4 Schematized apparatus for observing nuclear free induction decay. (a) Field geometry. (b) Electronics. (c) Rotating frame field and magnetization vectors.

signal caused by the rotating magnetization immediately after the 90° pulse. If the coil is of unit volume, cross-section area A, and has n turns, then the induced voltage will be

$$V_0 = -\frac{n}{c}\eta\frac{d\phi}{dt} = -\frac{1}{c}4\pi An\eta\frac{dM}{dt} = -\frac{1}{c}4\pi n\eta A\omega M_0 \qquad (2\text{-}38)$$

where ϕ is the flux linking the coil: $\phi = BA$, and $B = 4\pi M$, and η is a factor between zero and one, the "filling factor," which takes into account incomplete flux linkage between sample and coil. A problem at the end of the chapter will show that the magnitude of V can be as large as several millivolts for nuclear systems if the coil is part of a resonant circuit of reasonable Q.[5] Several millivolts is, of course, a very easily detectable signal at radio frequencies.

Equation (2-38) gives the signal V_0 immediately after the 90° pulse, where the full equilibrium magnetization M_0 is turned over into the transverse plane. The transverse magnetization, as we have seen, decays exponentially with a time constant T_2. Since $T_2 \gg 2\pi/\omega_0$, the signal is contained within a slowly varying envelope. The envelope decays according to the expression $V_0 \exp(-t/T_2)$, which is known as the "free induction decay."

Steady state solution

The next type of solution of the Bloch equations to investigate is the steady state solution in the presence of continuous H_1. We shall express the solutions of (2-35) in terms of the susceptibilities $\chi'(\omega)$ and $\chi''(\omega)$. To begin, we rewrite Eqs. (2-35) in component form in the rotating frame of the rf field, which, in the laboratory frame, is

$$\mathbf{H}_1(t) = \hat{\mathbf{i}}2H_1 \cos \omega t$$

The transformation is defined by $\omega = -\omega\hat{\mathbf{k}}$. We ignore the counter-rotating component, as in Chapter 1.

$$\frac{dM_z}{dt} = -\gamma M_y H_1 + \frac{M_0 - M_z}{T_1} \qquad (2\text{-}39a)$$

$$\frac{dM_x}{dt} = -\gamma M_y\left(H_0 - \frac{\omega}{\gamma}\right) - \frac{M_x}{T_2} \qquad (2\text{-}39b)$$

$$\frac{dM_y}{dt} = -\gamma\left[M_z H_1 - M_x\left(H_0 - \frac{\omega}{\gamma}\right)\right] - \frac{M_y}{T_2} \qquad (2\text{-}39c)$$

[5] The results of Eq. (2-38) will be in the Gaussian units, where V is expressed in statvolts. The result must be multiplied by 300 to express the results in volts.

In the steady state, the left-hand sides of Eqs. (2-39) vanish. From Eq. (2-39a) we see that $(M_0 - M_z)$ is proportional to $M_y H_1$. We expect, from the definitions of the susceptibilities χ' and χ'', that M_y will be proportional to H_1, so $(M_0 - M_z)$ is of the order H_1^2. As long as we restrict ourselves to terms linear in H_1, we can replace M_z by M_0 in Eqs. (2-39b) and (2-39c).

The algebra is simplified by introducing $\mathcal{M}_+ = M_x + iM_y$. Add (2-39b) to $\sqrt{-1} = i$ times (2-39c):

$$\frac{d\mathcal{M}_+}{dt} = -\mathcal{M}_+ \left[\frac{1}{T_1} + i\gamma\left(H_0 - \frac{\omega}{\gamma}\right)\right] + i\gamma M_0 H_1 \tag{2-40}$$

The steady state solution to Eq. (2-40) is

$$\mathcal{M}_+ = \frac{i\gamma M_0 H_1}{1/T_2 + i\gamma(H_0 - \omega/\gamma)} \tag{2-41}$$

We define $\omega_0 = \gamma H_0$, substitute $M_0 = \chi_0 H_0$, and then separate into real and imaginary parts to get

$$M_x = \chi_0 \omega_0 T_2 \frac{(\omega_0 - \omega)T_2}{1 + (\omega - \omega_0)^2 T_2{}^2} H_1 \tag{2-42a}$$

$$M_y = \chi_0 \omega_0 T_2 \frac{1}{1 + (\omega - \omega_0)^2 T_2{}^2} H_1 \tag{2-42b}$$

From the definitions of χ' and χ'', Eq. (2-30), we identify the components of the complex susceptibility:

$$\chi' = \frac{\chi_0 \omega_0 T_2}{2} \quad \frac{(\omega_0 - \omega)T_2}{1 + (\omega - \omega_0)^2 T_2{}^2} \tag{2-43a}$$

$$\chi'' = \frac{\chi_0 \omega_0 T_2}{2} \quad \frac{1}{1 + (\omega - \omega_0)^2 T_2{}^2} \tag{2-43b}$$

In Fig. 2-5, χ' and χ'' are plotted versus $(\omega_0 - \omega)T_2$.

The components of the complex susceptibilities have some interesting features. The full width at half-maximum of the absorption χ'' is $2/T_2$, and the peaks of the dispersion χ' are at $\omega = \omega_0 \pm 1/T_2$. Equations (2-43) are called Lorentz curves, after the line shape of the optical absorption and emission predicted by Lorentz using a damped simple harmonic

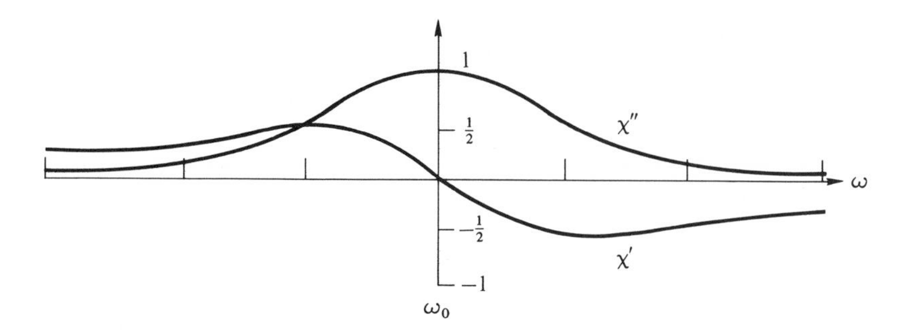

Fig. 2-5 The rf susceptibilities χ' and χ'' as a function of radio frequency ω. The vertical scale is in units of $\chi_0 \omega_0 T_2/2$; the horizontal axis is in units of $1/T_2$.

oscillator model of the atom. The power of the resonance technique is illustrated by the observation that the maximum value of χ'', $\chi_0 \omega_0 T_2/2$, may be written $\chi_0 H_0/\Delta H$, where $\Delta H = 2/\gamma T_2$ is the full width of χ'' in field units. In liquids, T_2 can be on the order of seconds, and hence the factor $\omega_0 T_2/2$ or $H_0/\Delta H$ can be as high as 10^8. Thus, although χ_0 might easily be 10^{-11}, and hence the static magnetization $\chi_0 H_0 = 10^{-7}$ G, the dynamic susceptibility at resonance may be 10^{-4}. Of course, the probing field H_1 is necessarily small (to be discussed), but the magnetization is at ω_0 rather than zero frequency, so the relatively simple techniques of rf signal amplification and detection are used.

Had we not neglected the difference between M_0 and M_z in Eq. (2-39c), the algebra leading to Eq. (2-43) would have been more complicated, but far from intractable. The result would have been to add the term $\gamma^2 H_1{}^2 T_1 T_2$ to the denominator of Eqs. (2-43a) and (2-43b), so that (2-43b), for example, would read

$$\chi'' = \frac{\chi_0 \omega_0 T_2}{2} \frac{1}{1 + (\omega - \omega_0)^2 T_2{}^2 + \gamma^2 H_1{}^2 T_1 T_2} \tag{2-44}$$

The term $S = \gamma^2 H_1{}^2 T_1 T_2$ is called the saturation factor. Regarded as a function of S, the absorbed power, $2H_1{}^2 \chi'' \omega_0$, is a maximum when $S = 1$. For $S \gg 1$, χ'' tends to zero; the resonance is said to be saturated.

The solution for M_z including saturation is also interesting:

$$M_z = \chi_0 H_0 \frac{1 + T_2{}^2(\omega_0 - \omega)^2}{1 + T_2{}^2(\omega_0 - \omega)^2 + \gamma^2 H_1{}^2 T_1 T_2} \tag{2-45}$$

It is worth displaying (and worth the student's time obtaining), because it can be used with previous general remarks to calculate W_{rf}.

Recall Eq. (2-27) for the excess population n:

$$n = \frac{n_0}{1 + 2W_{\mathrm{rf}} T_1} \tag{2-27}$$

Using $n_0 \gamma \hbar = 2\chi_0 H_0$, and $M_z = \frac{1}{2} n \gamma \hbar$, we equate (2-27) and (2-45) to obtain

$$W_{\mathrm{rf}} = \frac{1}{2} \frac{\gamma^2 {H_1}^2 T_2}{1 + {T_2}^2(\omega_0 - \omega)^2} \tag{2-46}$$

Note the rapid diminution of the transition probability as the rf frequency ω differs from ω_0. Of more interest is W_{rf} at $\omega = \omega_0$. To generalize as much as possible, we must consider the line shape for an absorption experiment. It is, of course, χ'', but we wish to express that shape as a function of $\nu = \omega/2\pi$, $g(\nu)$, which is normalized to unity:

$$\int_0^\infty g(\nu)\, d\nu = 1$$

The correct $g(\nu)$ for a Lorentz line is

$$g(\nu) = \frac{2T_2}{1 + 4\pi^2 {T_2}^2(\nu_0 - \nu)^2}$$

Value of $g(\nu)$ is $2T_2$ at $\nu = \nu_0$. Returning to Eq. (2-46), we then can substitute for T_2 the quantity $\frac{1}{2} g(\nu_0)$, and we get, for $\nu = \nu_0$,

$$W_{\mathrm{rf}} = \tfrac{1}{4} \gamma^2 {H_1}^2 g(\nu_0) \tag{2-47}$$

This formula is, in fact, the quantum mechanical result (the "golden rule") for the transition probability of a spin $\frac{1}{2}$ system, where the quantity $g(\nu_0)$ is the "density of final states," which explicitly expresses the width of the level via the shape and normalization. The quantity $\gamma^2 {H_1}^2/4$ is obtained from the square of the matrix element of the perturbation $-\mu_x H_1$ between the initial and final states:

$$|\langle \tfrac{1}{2} | \gamma H_1 I_x | -\tfrac{1}{2} \rangle|^2$$

Finally, to complete our earlier discussions, we can equate (2-33), $P = 2{H_1}^2 \chi'' \omega$, and (2-28), $P = n_0 \hbar \omega W_{\mathrm{rf}}/[1 + 2W_{\mathrm{rf}} T_1]$, and solve for χ'', which turns out to be (2-44). We conclude that the Bloch equations are consistent with our general discussions about energy absorption based on detailed balancing.

2-6. SOME EXPERIMENTAL CONSIDERATIONS

A great variety of experimental arrangements to observe nuclear magnetic resonance have been successfully used. Descriptions of some of the earlier ones are in the book by Andrew [4]. Although Andrew's book is old, most of the basic techniques still in use are described there. We want to analyze the experimental problem in a general way to establish the understanding with which the student can investigate on his own any particular circuit he may wish. Although our discussion will be couched in the language of radio frequencies—coils and capacitors are prominent—rather than microwave frequencies where resonant cavities are usually used, the analysis is really sufficiently general to apply to the microwave case also.

Q-meter detection

Consider a coil the inductance of which in the absence of a sample is L_0. If a sample of permeability μ occupies all space threaded by the magnetic lines of force generated by a current through the coil, then the inductance of the coil becomes

$$L = \mu L_0 = (1 + 4\pi\chi)L_0 \tag{2-48}$$

If the sample does not fill all space, the susceptibility χ must be multiplied by the filling factor η, as in Eq. (2-38). The quantity χ in Eq. (2-48) is the *complex* susceptibility defined in (2-31). Since no coil exists without resistance, unless it is made of superconducting wire, we must really always specify the coil's resistance R_0 as part of the coil. Our problem is to calculate the coil impedance $\mathscr{Z}_L = R_0 + i\omega\mathscr{L}$. Since $\mathscr{L}$ is complex, $i\omega\mathscr{L}$ has a real part (assume $\eta = 1$):

$$\begin{aligned}\mathscr{Z}_L &= R_0 + i\omega[1 + 4\pi(\chi' - i\chi'')]L_0 \\ &= R_0 + 4\pi\omega L_0\chi'' + i[1 + 4\pi\chi']\omega L_0\end{aligned}$$

To design an experimental method of detecting χ' and χ'', we must know their size relative to other terms in Eq. (2-49). Let us use, for numerical purposes, a fairly typical sample having a resonant frequency of 10 MHz in 10^4 G, and with a static susceptibility, χ, of 10^{-11}; $T_1 = T_2 = 10^{-3}$ sec. Then $4\pi\chi$ has a maximum value of 10^{-7}. A coil of wire that is useful at 10 MHz has an inductance of, say, a microhenry, so $\omega L_0 \simeq 60\ \Omega$, and a reasonable R_0 would be 1 Ω. Therefore, the reactive (imaginary) part of $\mathscr{Z}_L$ changes by one part in 10^7 as we go through resonance, and the resistive part changes by the fractional amount $4\pi\omega L_0\chi''/R_0 = 10^{-5}$. (It is convenient to identify the ratio $\omega L_0/R_0 = Q$ as the "quality factor." See a text on ac circuits if you are unfamiliar with the definitions and uses of this parameter in resonant circuits.) We conclude that a necessary feature of any circuit design is that it be particularly sensitive to small changes in the real or imaginary part of $\mathscr{Z}_L$.

With few exceptions, the coil L is used in a resonant circuit by placing a capacity C_0 in parallel with L such that the Larmor frequency, γH_0, is the same as the resonant frequency $\omega_0 = 1/(L_0 C_0)^{1/2}$. The simplest conceivable circuit for observing a nuclear resonance is the so-called "Q-meter" circuit, shown in Fig. 2-6. The oscillator and large resistor R

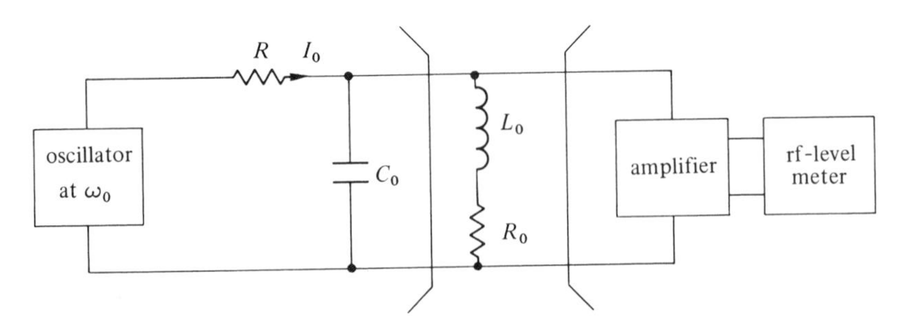

Fig. 2-6 Block diagram of Q-meter nuclear magnetic resonance detector.

form a constant current generator of current I_0. The parallel resonant circuit, tuned to ω_0, presents an impedance that, at resonance, is real: $Z_0 = Q\omega_0 L_0$. The voltage across it, $V_0 = I_0 Z_0$, is amplified, and its magnitude is detected by the peak-reading voltmeter. From Eq. (2-49), we see that as we go through resonance by changing H_0, for example, the real part of $\mathscr{Z}$ changes fractionally by the amount $4\pi\chi''Q$, so that the voltage at the amplifier input changes by

$$\Delta\mathscr{V} = I_0 4\pi\chi'' Q^2 \omega L_0$$

or by the fractional amount

$$\frac{\Delta\mathscr{V}}{V_0} = \frac{I_0 4\pi\chi'' Q^2 \omega L_0}{I_0 Q\omega L_0} = 4\pi\chi'' Q \tag{2-49}$$

We shall now give a word of explanation as to why we were able to ignore the change in reactance of the resonant circuit. For future uses it is best explained with a diagram, rather than analytically. Figure 2-7 is a complex plane diagram, sometimes called a "phasor" diagram, in which the magnitude and phase of the off resonance voltage across the sample is

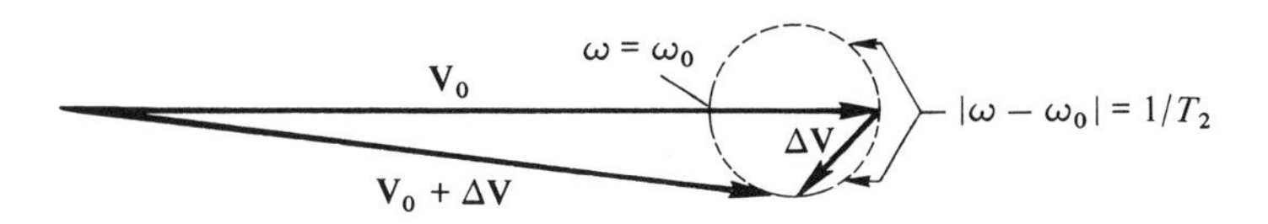

Fig. 2-7 Magnitude and relative phase diagram for voltages in Q-meter detection.

given by V_0, the "signal" voltage by $\Delta\mathscr{V}$, and the total voltage by $V_0 + \Delta\mathscr{V}$, the *magnitude* of which is detected by the peak-reading voltmeter. The complex signal $\Delta\mathscr{V}$ is given by

$$\Delta\mathscr{V} = -V_0(4\pi\chi'' + i4\pi\chi')Q \tag{2-50}$$

Reference to Eq. (2-50) and Fig. 2-7 shows that only the component of $\Delta\mathscr{V}$ in phase with V_0 is effective in determining the *length* of $V_0 + \Delta\mathscr{V}$. Since V_0 is real, because $Z_0 = Q\omega_0 L_0$ is real, the Q-meter circuit detects only χ'' to first order. The imaginary component, $-iV_0 Q4\pi\chi'$, is said to be in quadrature and affects the length only to second order in χ'.

The analysis of the "series-parallel" resonant circuit of Fig. 2-8a is

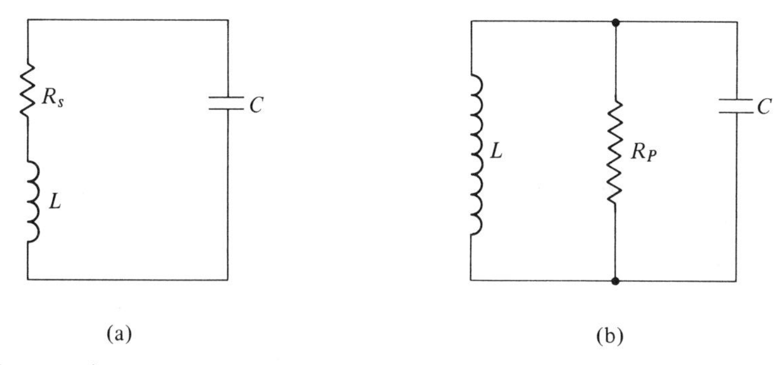

Fig. 2-8 (a) Series-parallel tank circuit. (b) Equivalent parallel circuit.

algebraically clumsy. Equation (2-50) is an approximation not only to first order in χ, but also to the extent that $\omega_0 = 1/(L_0 C_0)^{1/2}$ is the resonant frequency only in the approximation $Q \gg 1$. It is more convenient to analyze the *admittance* $\mathscr{Y}$ of the equivalent parallel circuit, Fig. 2-8b. The fictitious resistance R_P is related to R_s, in the high Q approximation, by $Q = \omega L/R_s = R_P/\omega L$. Then it is easy to show that $\mathscr{Y} = 1/R_P(1 + 4\pi\chi Q)$, and, by virtue of the smallness of χ, $\mathscr{Z} = 1/\mathscr{Y} = R_p(1 - 4\pi\chi Q)$, when $\omega_0 = 1/(L_0 C_0)^{1/2}$—hence Eq. (2-50).

The discussion of the "Q-meter" detector, so named because the detected voltage is proportional to the change in Q of the circuit, hence to χ'', was phrased to make it clear that the working of the circuit depends on the *selection* of the component of the signal in phase with a much larger voltage, in this case V_0. A clear grasp of this concept is necessary to appreciate the operation of bridges in magnetic resonance detection.

Figure 2-7 and the subsequent discussion make the point that the Q-meter circuit picks out χ'' because V_0 and $4\pi Q\chi'' V_0 = \text{Re}(\Delta\mathscr{V})$ are in phase. A large number of bridge circuits have been devised to accomplish this purpose, both in the microwave and rf regions. The archetype of the rf phase reference or coherent detection technique can be summarized in the block diagram of Fig. 2-9. In the discussion, it is understood that in

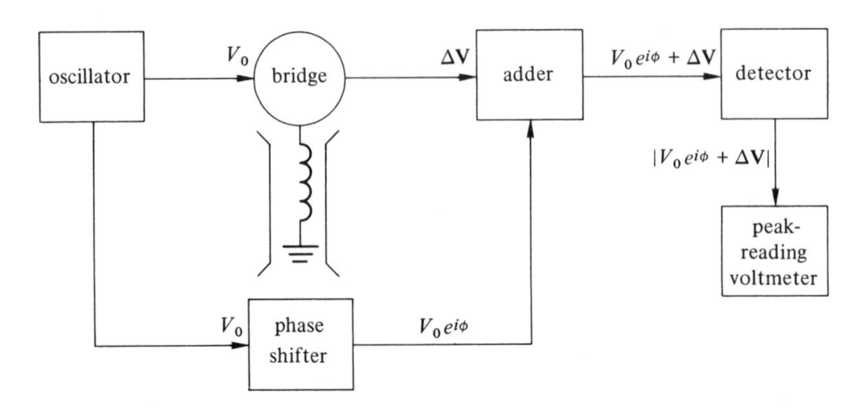

Fig. 2-9 Schematic archetypical bridge circuit.

the output of the oscillator $V_0 e^{i\omega t}$, the $e^{i\omega t}$ is suppressed. We assume V_0 is real for simplicity. The "bridge" is a device which, when balanced, has zero output, and which is unbalanced by the change in sample circuit impedance on resonance. The only output is the signal voltage $\Delta\mathscr{V}$. The phase shifter in Fig. 2-9 shifts the phase of the oscillator voltage by ϕ and supplies a reference signal to the adder, whose output is $V_0 e^{i\phi} + \Delta\mathscr{V}$. The detector is just the "peak-reading voltmeter" as before, and the amplitude $|V_0 e^{i\phi} + \Delta\mathscr{V}|$ is, to first order in $\Delta\mathscr{V}$, just V_0 plus the component of $\Delta\mathscr{V}$ in phase with the reference signal. One may write down the result more easily by pretending we shifted the phase of $\Delta\mathscr{V}$ by $(-\phi)$ instead of the phase of V_0 by $(+\phi)$ and by computing the real part of $\Delta\mathscr{V} e^{-i\phi}$:

$$\text{Re}(\Delta\mathscr{V} e^{-i\phi}) = 4\pi V_0 Q\chi'' \cos\phi - 4\pi V_0 Q\chi' \sin\phi \qquad (2\text{-}51)$$

Choosing $\phi = 0$ with the phase shifter gives χ'', $\phi = 90°$ chooses χ', and, as advertised, any admixture may be chosen as well.

In real experimental arrangements, the functions of two or more of the separate components of Fig. 2-9 are usually combined into one device. If a real rf bridge is used in place of the box marked "bridge" in Fig. 2-9, it may be balanced by dividing V_0 and sending half of it through an arm, after which it recombines with the half that went through the signal arm with, of course, equal amplitude and opposite phase. If the bridge is operated in this fashion, completely balanced (off resonance, anyway), then the reference part of Fig. 2-9 is needed. Often, the bridge is unbalanced either in amplitude or phase (or a combination of the two), so that the steady unbalance signal is much larger than the signal. Then the external reference channel in Fig. 2-9 is unnecessary, and the unbalanced bridge has also performed the function of the adder. On the other hand, if the bridge is balanced, the "adder" and detector may be combined, as is often done in modern instrumentation, by using a device called a "mixer," which takes two signals applied to two inputs and puts out of the third port the sum and difference frequencies. In our case, the difference frequency is zero, and its magnitude is given by Eq. (2-51).

Every statement previously made for rf techniques (frequencies up to, say, 100 MHz) has its equivalent in microwave techniques. The tuned LRC circuit is replaced by a resonant cavity, which has a "Q" and to which the language of impedances, voltages, and currents may also be applied. Microwave bridges are in many ways easier to understand than the older rf circuits, and in recent years there has become available an rf version of the time-honored component of the microwave bridge, the hybrid junction or "magic tee." It is now possible to make a nuclear resonance spectrometer that is an exact copy of the microwave equivalent, even to the extent of the language used to describe it.

All the original nuclear resonance by the Harvard and Stanford groups used bridge techniques. The Stanford group lead by Bloch used a device, the crossed coil spectrometer, which is perhaps the instrument of choice if space around the sample permits. A sketch of the essentials of the arrangement is shown in Fig. 2-10. Unlike other methods the bridge balance is achieved by geometry. The field H_1 is applied in the y direction in the laboratory by the split coil called the transmitter coil, and the signal is induced in the orthogonal coil, the receiver coil, in the x direction of the laboratory frame. Remember that the induced magnetization rotates in the xy plane; thus either coil sees a magnetization varying at frequency ω. The bridge is balanced because, in principle, the lines of flux from the transmitter do not link any receiver coil turns. The unbalance is achieved by the natural lack of complete orthogonality, and control of the unbalance can also be achieved by mechanical means with "paddles," devices that "steer" the flux by means of currents induced in them by the field. Since the balance is achieved mechanically, it is somewhat broad band, and the

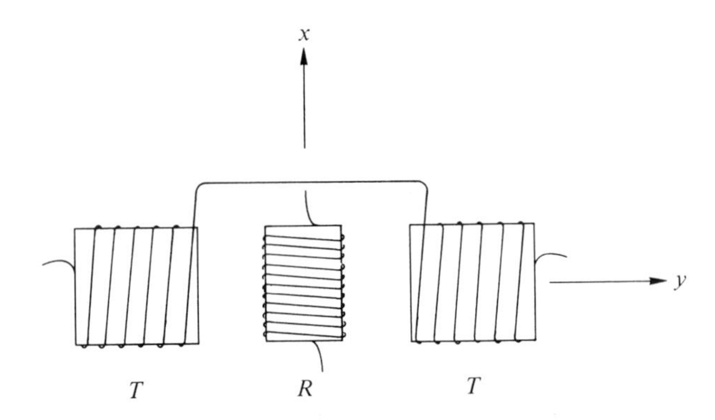

Fig. 2-10 Crossed coil geometry: T is the split transmitter coil and R the receiver coil containing the sample; H_0 is applied perpendicular to the page. Not shown are flux steering arrangements (paddles).

resonant frequencies of the transmitter and receiver can be swept synchronously if necessary. One can also balance as well as possible mechanically and achieve the rest of the balance and the reference through an electronic phase shifter, as in Fig. 2-9. There has even been a microwave version of the crossed coil spectrometer, with two resonant cavities coupled by the sample.

One somewhat academic difference exists between what is measured by the crossed coil apparatus and the others we have discussed. Strictly speaking, the complex susceptibility χ is a tensor rather than a scalar quantity, and the single coil measures the response M_x to a field applied in the x direction: we should write $M_x = \chi_{xx} H_{1x}$. The crossed coil apparatus measures χ_{xy}, since the receiver coil is in the x direction and the transmitter coil is in the y direction: $M_x = \chi_{xy} H_{1y}$. There is no case in pure nuclear magnetic resonance in which the distinction is important, since the precessing magnetization is circularly polarized. In the case of pure quadrupole resonance (see Chapter 4), the magnetization is actually linearly polarized, and $\chi_{xy} = 0$; the crossed coil technique does not work!

Marginal oscillators

Bridge and Q-meter circuits separate the function of oscillator and receiver. The separation of these functions has the advantage that the oscillator contribution to the noise of the device can be made negligible, so that the entire noise generation is in the receiver and sample circuits. The separation of these functions has the disadvantage that it is awkward to sweep the frequency rather than the external field in displaying the resonance or in searching for it. That objection is less true of the crossed coil circuit, previously mentioned, because the bridge balance depends on geometry rather than on a number of frequency-sensitive circuit elements. Still, it is quite awkward to sweep over a factor of three in frequency unless the generator and receiver functions are combined. That object is achieved if the sample tank circuit is made a part of the oscillator tank. The device is then operated so that the voltage level of oscillator depends critically on the tank circuit Q. Although the term $4\pi\chi' Q$ also affects the frequency of oscillation, normally only the level of operation is monitored, so the circuit detects χ''. The circuits thus used are at least a factor of two less sensitive than bridge circuits (because of the contribution of oscillator noise), and they become very poor at low rf levels, where the noise properties of the combined oscillator-receiver system become poor. But the convenience of being able to sweep frequency by adjusting only the capacitor in the tank circuit has made these circuits very popular. Robinson has described a clever combination of Q-meter and marginal oscillator circuits that seems to have all the advantages of Q-meter and marginal oscillator circuits, plus the ability to operate with very low rf levels (to 100 μV) on the sample coil (Robinson [5]).

Detection techniques

We have progressed gently in this chapter from elementary statistical mechanics increasingly toward experimental considerations. This is as it should be, since the numerical magnitudes of magnetization, the dynamical behavior of that magnetization as it interacts with the externally applied fields with the lattice, and with itself (T_2), all force onto the experimentalist the techniques he uses. To conclude this chapter, we examine some of the methods commonly used—and used not only in magnetic resonance—to display with greater clarity, with as much freedom from noise interference as possible, the magnetic resonance signal. A few words about noise in general will have to suffice to establish the motivation for the techniques used; for even a modest step beyond mere introduction to the theory and practice of noise on electronic signals, the student will have to consult a text on the subject (particularly recommended is the book by Robinson [5]).

In discussing the various nuclear resonance techniques, we have sloughed over a number of crucial points in the interest of moving the narrative along. Let us remedy the fault by a rhetorical question: What are the major sources of noise in typical magnetic resonance experiments? The answer is oscillator noise, source noise, receiver noise, detector noise, and microphonics. The marginal oscillator circuits are particularly susceptible to oscillator noise; the bridge circuits and Q-meter circuits may be operated so that oscillator noise is not a factor. Receiver noise is the best kind to have limiting your experiment because it can be combatted by application of design skill and/or money, and it will always be a limitation until it is smaller than the inherent noise of the source. The detector, the device that converts the radiofrequency into direct current, has one important property. It is usually a diode, so it converts to direct current efficiently only when the radiofrequency applied to it is, say, greater than $\frac{1}{2}$ V. That is another rather good reason for applying a sizable reference radiofrequency in addition to the large one required by our analysis of phase detection.

Now, if the device labeled "peak-reading voltmeter" in Fig. 2-9 is actually a dc instrument, one is at the mercy of the instabilities of dc amplifiers, since very frequently rf amplification preceding detection is insufficient to provide a signal of convenient size. The problem is overcome by modulating the signal at an audiofrequency ν_m, so that the detected signal is at frequency ν_m, not zero, and may be amplified further. If the signal is somehow modulated, the detected signal is at ν_m and noise near ν_m is of importance. Diodes contribute noise with a $1/\nu$ frequency spectrum down to quite low frequencies. This restriction is particularly important for microwave diode detectors. Microphonic noise may, in principle, be overcome by good experimental design, but the ideal is often hard to achieve in practice. Those with long experience with microphonics seem to agree that a noise spectrum of $1/\nu$ may not be bad approximation in practice.

The source noise remains. A parallel tuned circuit on resonance looks like a resistance $R_s = Q\omega_0 L_0$. The unavoidable noise presented to the input terminals of the receiver is called Johnson noise, and is given by the Nyquist formula

$$V_N{}^2 = 4R_s k_B T \,\Delta\nu \tag{2-52}$$

where the noise resistance V_N is in volts, R_s is in ohms, T is in degrees Kelvin, k_B, Boltzmann's constant, is 1.38×10^{-23} J/K, and $\Delta\nu$, the bandwidth, is in sec^{-1}. Note that Eq. (2-52) depends on $\Delta\nu$, but is independent of ν, the center frequency.

Put together all these factors and one concludes that the game to play is to modulate the signal at as high a frequency as possible and to use as narrow a bandwidth as possible, dictated by Eq. (2-52). Figure 2-11 shows schematically the most common method for impressing a modulation on a signal. The ordinate is χ'' and the abscissa is H_0, which is varied around the mean value by an amount H_m, at frequency ν_m. Thus, $H(t) = H_0 + H_m \cos 2\pi\nu_m t$, say. If H_m is smaller than $1/\gamma T_2$, the rf signal at the detector will be $V(t) = (V_0 + V_m \cos \omega_m t) \cos \omega_0 t$, where V_m is proportional to $d\chi''/dH_0$. The last clause follows from Fig. 2-11, and would be

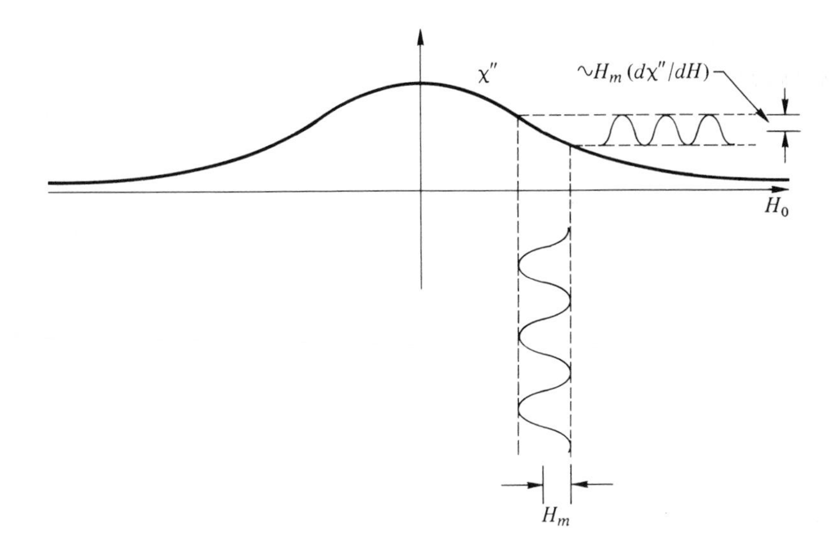

Fig. 2-11 Graph to show that lock-in detects $d\chi''/dH$.

exact if χ'' were exactly straight between $H_0 - H_m$ and $H_0 + H_m$: the output signal $V(t)$ is amplitude modulated with a modulation depth proportional to the derivative of χ'', at least to first order in an expansion parameter $\gamma T_2 H_m$. The frequency spectrum of the amplitude modulated rf signal has a main peak at ν_0 and sidebands at $\nu_0 \pm \nu_m$. The signal power is in the sidebands. After detection (now we had better say after the *first* detector), the signal power is at a frequency ν_m. ν_m is usually a low audio frequency for nuclear resonance, but may be as high as 100 kHz in some commerical electron spin resonance spectrometers. ν_m must satisfy the requirements $\nu_m \ll 1/T_2$, so that the magnetization can follow the field and always be in the steady state corresponding to our solutions of the Bloch equations.

The ultimate signal we wish to record is to be at direct current, that is, at $\omega = 0$. Clearly the job can be accomplished by the same method by which the rf signal was rectified, and, for much the same reasons, a phase-sensitive detector is employed. Since the phase of the audio signal carrying the signal power is known, phase-sensitive detection at the audio frequency may be used to discriminate against noise at ν_m not in phase with the signal at ν_m. The device that accomplishes this task, the second detector, is often called a lock-in. The terminology originated in radar work during World War II.

Finally, the bandwidth $\Delta\nu$ is normally limited not by the bandwidth of an rf amplifier at ν_0, or the audio amplifier at ν_m, but by an RC network arranged as a simple "integrator" at the output of the lock-in detector.

Figure 2-12 shows $d\chi''/dH_0$ of the ^{27}Al resonance in aluminum metal, taken with the bridge, rf amplifier, and modulation lock-in technique described in this chapter.

There are many circumstances in which the signal is strong enough that signal modulation is neither desirable nor even possible. The only thing to be done is to sweep through the resonance with the external field fairly rapidly, making sure the subsequent amplifiers and detectors have enough bandwidth to reproduce the signal faithfully. During this rapid sweep, it may easily turn out that dH_0/dt may be too large to be ignored in the Bloch equations. Their solution then becomes much more difficult than our steady state solution; some of the details were worked out in the first investigation by Bloch *et al.*, since their experiments were done that way. Figure 2-13 shows a proton resonance in water. A problem at chapter's end deals with some of the aspects of that signal.

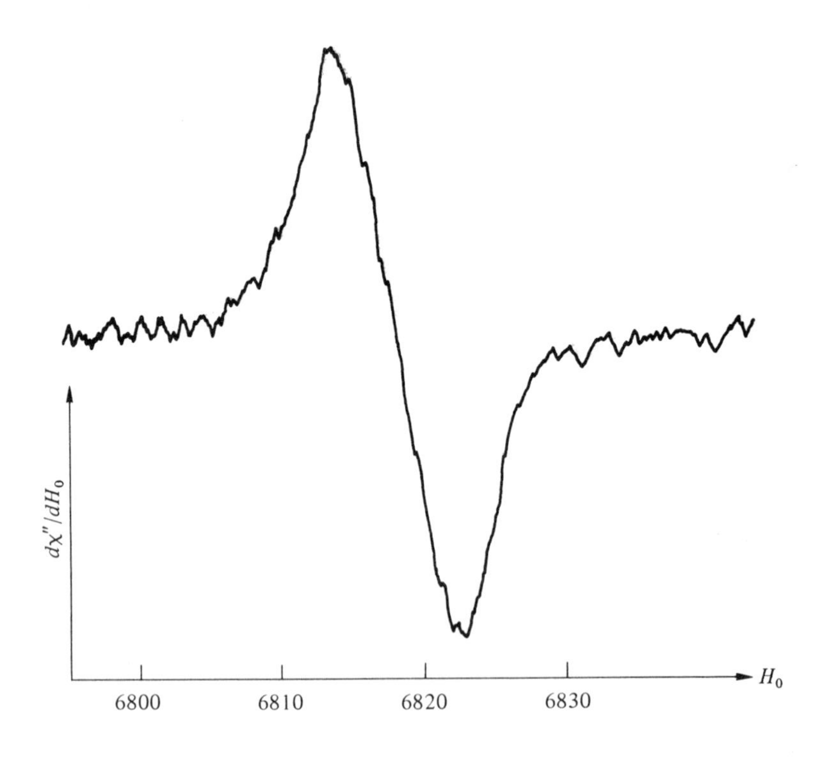

Fig. 2-12 Nuclear magnetic resonance signal, $d\chi''/dH$, of ^{27}Al in Al metal.

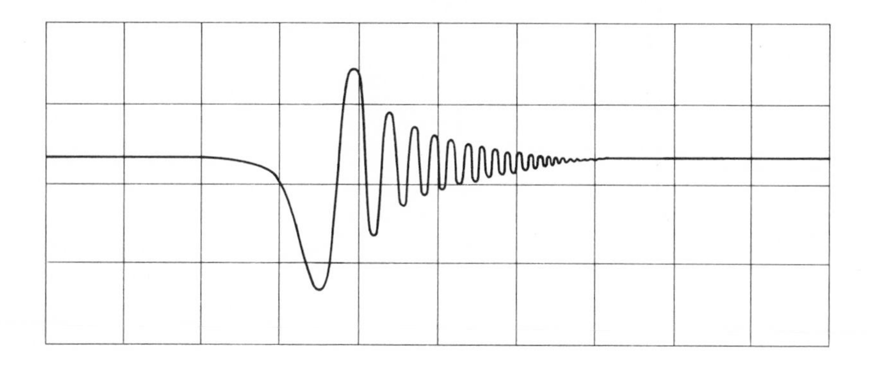

Fig. 2-13 Proton resonance in water, showing the phenomenon of "wiggles" (after Abragam [2], Figure III, 15).

Spin echoes

We conclude this chapter on some of the macroscopic aspects of magnetic resonance, and on the Bloch equations, by describing a phenomenon that appears, at first glance, to be too special, too clever a trick, to merit description in a text with an aim as general as ours. But the phenomenon of the spin echo, of a physical system emitting spontaneous signals if suitably prepared, has been exhibited in a sufficient variety of physical systems to make clear that it is a general property of systems with the sort of nonlinearity exhibited by the Bloch equations. The physical systems in which the spin echo has been seen, beyond the original nuclear magnetization system, include electron spin systems, atomic systems with induced electric dipole moments (at optical frequencies—the "photon echo"), and plasmas.

The spin echo is most easily seen in the following sort of system. Let T_2 and T_1 be rather long, but let T_2 be the parameter characterizing internal spin-spin interactions. Place the sample in a somewhat inhomogeneous external field $H_0 + H(r)$, so the variation of the external field over the sample, described by $H(r)$, can be described by a $T_0^* \ll T_2$. The coordinate r ranges over the sample. The following sequence of rf pulses is applied to the sample: (a) at $t = 0$, a 90° pulse; (b) at $t = \tau$, a 180° pulse, where $T_2 > \tau > T_2^*$. The result is a spontaneous signal of magnitude $V_0 \exp(-2\tau/T_2)$ appears at $t = 2\tau$.

Examine the process with the aid of Fig. 2-14. The 90° pulse tips M_0 into the x direction of the coordinate system rotating at $\omega_0 = \gamma H_0$ (see Fig. 2-14a). The "isochromats" precess in both directions relative to the x axis until they have fanned out completely, in $t \gtrsim T_2^* \simeq 1/\gamma\overline{H(r)}$, where $\overline{H(r)}$ is some average deviation of the field from H_0 (2-13b). Between the

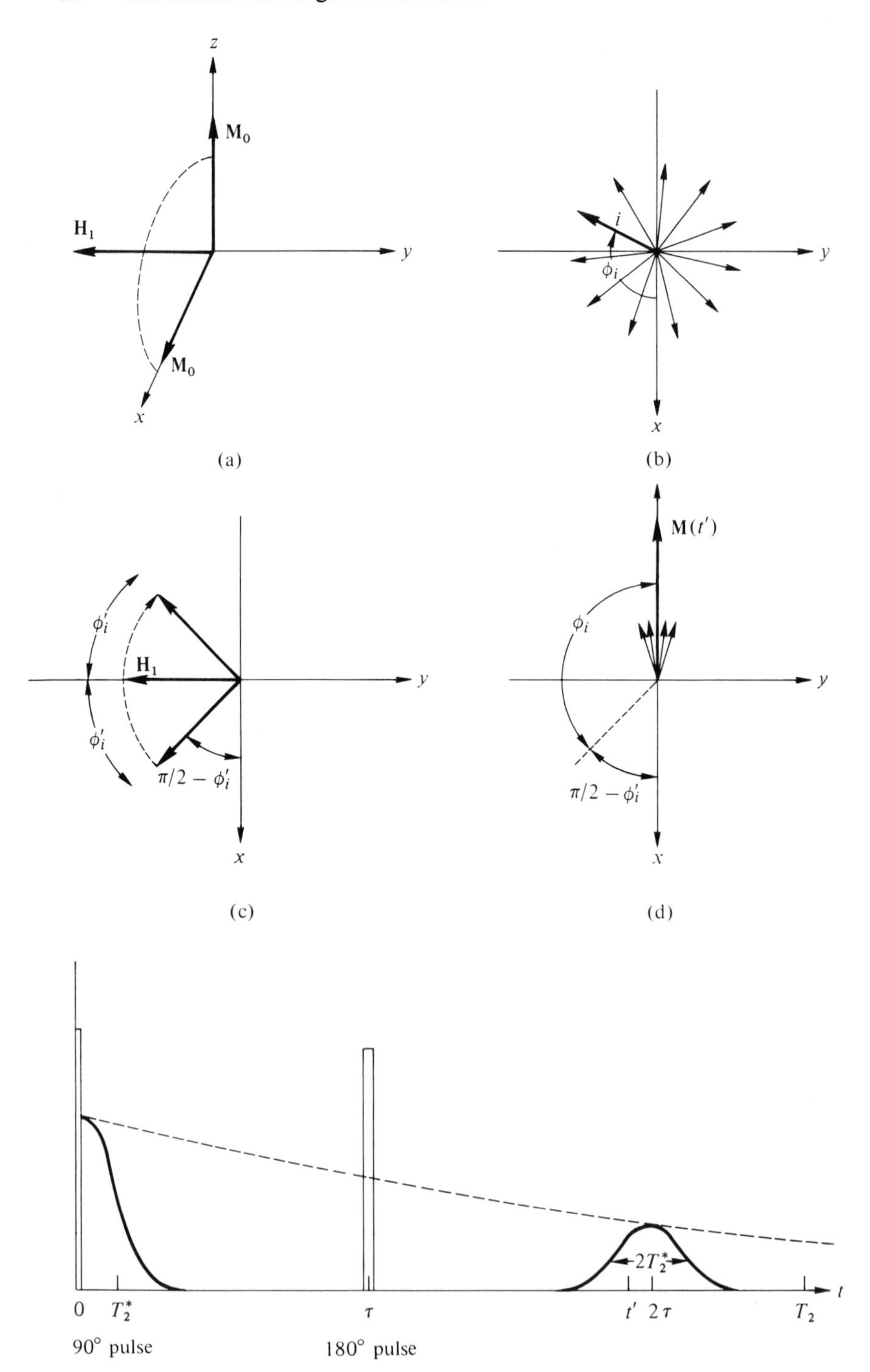
z
$\mathbf{M}_0$
$\mathbf{H}_1$
y
$\mathbf{M}_0$
x
(a)
i
ϕ_i
y
x
(b)
ϕ_i
$\mathbf{H}_1$
y
ϕ_i'
$\pi/2 - \phi_i'$
x
(c)
$\mathbf{M}(t')$
ϕ_i
y
$\pi/2 - \phi_i'$
x
(d)
$2T_2^*$
t
0
T_2^*
τ
t'
2τ
T_2
90° pulse
180° pulse
(e)

end of the 90° pulse and the beginning of the 180° pulse at $t = \tau$, a representative magnetization vector will precess relative to the x axis of the rotating frame by the angle $\phi_i = \gamma H(r_i)\tau$. The index i labels a particular small region of the sample. It is convenient to label the precession angle relative to the y axis: $\phi_i = (\pi/2) + \phi_i'$, where ϕ_i' is illustrated in Fig. 2-14b, for time t just before the 180° pulse. The 180° pulse flips the entire "pancake" about the y axis. The position of the magnetization that had precessed ϕ_i is shown in Fig. (2-14c). It is now at $\phi_i = (\pi/2) - \phi_i'$. It continues to precess in the same sense during the subsequent time, so that, at 2τ, it has accumulated another $\phi_i = (\pi/2) + \phi_i'$. The total phase accumulated at 2τ, including the 180° pulse, is thus $(\pi/2) - \phi_i' + (\pi/2) + \phi_i' = \pi$. The accumulated angle is independent of the labeling index i; therefore, all regions of the sample contribute to a signal at $t = 2\tau$, and all magnetization vectors add to form a macroscopic vector in the x direction in the rotating frame. The shape of the echo may be described as two free induction decays back-to-back, as shown in Fig. 2-14e. The signal is attenuated from the initial magnitude of the free induction decay, V_0, by *real* T_2 processes, which result in unrecoverable loss of phase coherence, as distinct from losses caused by static magnetic field inhomogeneities. If the echo height is measured as a function of τ, the 90° to 180° pulse separation, the echo height will follow the exponential $\exp(-2\tau/T_2)$.

The preceding example of a spin echo is the barest introduction. Many variations in pulse length, sequence, and number are possible. The technique is frequently used in the study of liquids and solids, where various contributions to the line widths can often be unraveled. There are many modern applications of the basic idea in seemingly remote fields, such as nonlinear optics and plasma diagnostics.

2-7. CONCLUSION AND LITERATURE SURVEY

Most of the material in this chapter is covered in every text and review article that discusses magnetic resonance. The material requiring statistical mechanics may be found in Feynman [6] and Reif [7], and, of course, all more advanced treatments of statistical mechanics. Treatments of the Bloch equations more or less on a somewhat more advanced level than found here are in the article by Pake [8], the books by Andrew [4], Kopfermann [9], Slichter [3], and Abragam [2]. Discussions of experimental

---

Fig. 2-14 (a) Effect of the 90° pulse on the magnetization vector $\mathbf{M}_0$. (b) "Pancake" formed in xy plane of rotating frame after $t > T_2^*$. (c) Effect of the 180° pulse on the ith magnetization vector. (d) Precessing magnetization vectors as echo is forming, corresponding to t' of (e). (e) Pulse sequence and signals seen at various times. t' corresponds to vector diagram of (d). Dotted line traces echo envelope as τ is varied.

techniques may be found in Andrew and Kopfermann. A totally exhaustive compendium of everything done until 1966 in the area of microwave instrumentation is in the book by Poole [10]. The previously mentioned monograph by Robinson [5] on noise in rf circuits gives an excellent and clear discussion of elementary principles, with applications in the last chapter to magnetic resonance instrumentation.

We should perhaps conclude by emphasizing that much of our discussion in this chapter is applicable beyond nuclear magnetic resonance, even though the language of NMR was used for convenience. The Bloch equations are not valid generally for nuclear or electron spins in solids (except conduction electrons in metals), but they are correct for liquids, and do serve to provide an introduction to the phenomena involved. It is also useful pedagogically to have them available for calculating χ' and χ'', in order to provide concrete examples of these important but slightly abstract functions. We remind the student that the chapter has been macroscopic—we moved as quickly as possible to macroscopic magnetizations and response functions for macroscopic samples. The "model theory" we used, the Bloch equations, was entirely phenomenological, and also dealt only with macroscopic magnetizations and fields. The observed signals are also macroscopic voltages, and the experimental problems of measuring them formed an important part of the chapter.

Problems

2-1. Find the correct approximation to Eq. (2-17) in the limit $\gamma\hbar H_0/2kT \ll 1$. What temperature T must be reached for protons ($\gamma = 2.6 \times 10^4$ G^{-1} sec^{-1}) in a field of 10^4 G before the high temperature approximation is wrong by 10%? Find the same quantity if the magnetic moment is that of the electron ($\gamma_e = 1.74 \times 10^7$ G^{-1} sec^{-1}).

2-2. Obtain Eq. (2-23) directly from Eq. (2-17) in the high temperature limit of problem 2-1. Find χ_0 for protons in water at room temperature.

2-3. From the Bloch equations and Eq. (2-33), obtain the maximum power absorbed per unit volume from the protons in water at 60 MHz. Use an H_1 that makes the saturation parameter $S = \gamma^2 H_1{}^2 T_1 T_2 = 1$ in Eq. (2-44). Assume $T_1 = T_2 = 3$ sec.

2-4. Find the approximate maximum voltage at the input of an rf receiver produced by the free induction decay after a 90° pulse applied to $Na^{23}Cl$. Assume a receiver coil of 5 turns, cross section 1 cm^2, and a resonant frequency of 10 MHz.

2-5. Find the signal at the receiver input from a spin echo in water under the following experimental conditions: coil, 10 turns, area 1 cm^2; pulses, 90 to 180° sequence, $\tau = 2$ sec, $T_2 = 3$ sec.

2-6. Invent other pulse sequences involving more than two pulses that give rise to other echoes. (There are almost limitless possibilities.)

2-7. Figure 2-13 shows the proton resonance in H_2O in a relatively homogeneous field. The beating phenomenon, known as "wiggles," arises because the external magnetic field is changing rapidly enough that the field is off resonance before the transverse magnetization has decayed. Since the magnetization precesses at a frequency proportional to the field, the signal beats with the constant oscillator frequency, and the constantly changing difference frequency appears in the detected signal as "wiggles." The sweep in Fig. 2-13 is linear at 0.1 sec and 5 mG per division. Estimate T_2^* from the figure and the field inhomogeneity at the sample.

2-8. The cw (continuous wave) or steady state resonance signal in a particular liquid sample consists of two nearby lines of equal intensity and transverse relaxation time T_2. Calculate the free induction decay following a 90° pulse if a transient experiment is performed. If the lines are separated by angular frequency $\Delta\omega$ in the cw experiment, show that an approximately 90° pulse can be applied to both with a single pulse applied at a frequency midway between the lines if the 90° pulse condition is satisfied, and the pulse length satisfies the inequality $\tau \ll (\Delta\omega)^{-1}$. (Satisfying the 90° pulse condition simultaneously means $H_1 \gg \Delta\omega/\gamma$, where $\Delta\omega/\gamma$ is the separation of the lines in field units. Such an H_1 is said to be sufficient to "cover" the lines.)

2-9. Discuss the shape of the echo formed after two pulses of problem 2-8.

References

1. G. Pake, *Paramagnetic Resonance*, W. A. Benjamin, Inc., New York (1962).
2. A. Abragam, *Principles of Nuclear Magnetism*, Oxford University Press, London (1961).
3. C. P. Slichter, *Principles of Magnetic Resonance*, Harper & Row, New York (1963).
4. E. R. Andrew, *Nuclear Magnetic Resonance*, Cambridge University Press, Cambridge, England (1955).
5. F. N. H. Robinson, *Noise in Electrical Circuits*, Oxford University Press, London (1962).
6. R. P. Feynman, R. B. Leighton, and Matthew Sands, *The Feynman Lectures on Physics*, Addison-Wesley Publishing Co., Reading, Massachusetts, vols. 1, 2, 3 (1965).
7. F. Reif, *Statistical Physics*, McGraw-Hill Book Company, New York (1967).
8. G. Pake, "Nuclear Magnetic Resonance," *Solid State Physics*, F. Seitz and D. Turnbull, Eds., Academic Press Inc., New York, vol. 2 (1956), pp. 1–91.
9. H. Kopfermann, *Nuclear Moments*, Academic Press Inc., New York (1958).
10. C. Poole, *Electron Spin Resonance; A Comprehensive Treatise on Experimental Techniques*, Interscience Publishers, New York (1967).

CHAPTER 3

Line Widths and Spin-Lattice Relaxation in the Presence of Motion of Spins

This chapter will be concerned with providing a microscopic model with which to estimate the parameters of the Bloch equations T_1 and T_2. To do so, we shall introduce the complementary concepts of random frequency modulation and random walk. The main use of these concepts will be to achieve a clear understanding of the motional narrowing of nuclear magnetic resonance lines, but the ideas are also important in the understanding of other experiments in modern physics, which we shall discuss in the latter part of the chapter.

3-1. INTRODUCTION

From a strictly economic point of view, magnetic resonance owes a great deal to its usefulness in chemistry. It is useful in chemistry because the nuclear magnetic resonance line widths in liquids are so narrow that resonant frequencies differing by as little as a part in 10^8 may often be resolved. It has been found that the frequency of nuclei in different chemical surroundings depends on the details of the surroundings. For example, the resonance frequencies of protons in ethyl alcohol, CH_3CH_2OH, are divided into three groups, corresponding to the protons in CH_3, in CH_2, and in OH. The *chemical shifts* caused by the different chemical environments are small, however, compared with the magnetic dipole-dipole interaction between the protons in the molecule, which corresponds to a magnetic field of 10 G produced on one proton by another an angstrom away. Our first guess ought to be that the resonance lines would be about 10 G wide, precluding a resolution of better than one part in 10^4. We shall be concerned with the solution to this apparent paradox in this chapter.

To complete and sharpen the paradox, we write down the magnetic field produced at a distance r by a point magnetic dipole.

$$\mathbf{H}_d = -\frac{\boldsymbol{\mu}}{r^3} + 3\,\frac{\boldsymbol{\mu}\cdot\mathbf{r}}{r^5}\,\mathbf{r} \tag{3-1}$$

The field $\mathbf{H}_d$ has the familiar dipole shape; the interaction energy of one dipole at the origin with another dipole at $\mathbf{r}$ has a rather complicated angular dependence, a $1/r^3$ radial dependence, and it depends as well on the relative orientation of the dipoles. Thus the dipolar field varies from site to site, and cannot be exactly the same for each nucleus. The simple estimate of $|\mathbf{H}_d| \simeq 10$ G between protons, for example, is calculated by using μ/r^3, where r is the nearest neighbor distance, and is to be taken as an estimate of the rough magnitude of local fields in hydrogenous solids. Thus we present the paradox as follows. The widths of nuclear resonance lines in a liquid can be a fraction of a cycle per second in the presence of local dipolar interactions as large as 50 kHz.

It is our purpose in this chapter to resolve the paradox both quantitatively and qualitatively, to discuss the Bloch equation relaxation times T_1 and T_2 as a function of the resonance field or frequency, and to broaden the discussion to include the phenomena of "exchange narrowing" and "exchange broadening" in nuclear and electron paramagnetic resonance, the Mössbauer effect, and the intensity of Bragg reflections in X-ray crystallography. We also hope to make clear the distinction between "motional narrowing" (a name for the effect we want to discuss) and the "pressure broadening" of lines in atomic spectra. It must be admitted at the outset that the way of understanding the phenomena that we shall develop is a natural one for magnetic resonance but not so natural for some of the other phenomena. Nevertheless, it is important to grasp phenomena from as many viewpoints as possible, so it is worth the strain placed on our method to do so.

The resolution of the paradox involves the recognition that in a liquid, a gas, and even in solids under some circumstances, the resonant spins can move substantial distances relative to their average spacing in T_2, and even in a Larmor period in some circumstances. Thus, the local fields with which we are dealing are not static but rather time dependent, most often in a random way, and we must develop ways to understand how the time dependence affects the resonance experiment. To begin with, we shall treat the problem temporally; that is, we shall examine the precession of a typical spin as a function of time. Although we shall thus provide ourselves with a useful formula with which we can estimate T_2 in a wide variety of cases, we shall not have grasped the significance of the term "motional narrowing" until we have reexamined the problem in frequency space. To do that, we use some of the concepts and language of frequency modulation of a classical oscillator.

The language will be classical throughout, and we shall deal in qualitative estimates. Such an approach does the subject something of an injustice, since the phenomena are susceptible to quite precise and elegant formulation via the density matrix of quantum statistical mechanics. We henceforth banish from these pages any serious reference to the density matrix, and guide more ambitious and sophisticated readers to the text by Slichter [1].

3-2. RANDOM WALK CALCULATION OF T_2

We imagine ourselves sitting on a proton in water, for example, responding to the various magnetic fields in the sample. The strongest of these is the external field H_0, and we can dispose of it by looking at the world from a reference frame rotating at $\gamma H_0 = \omega_0$. The internal fields caused by the magnetic dipoles of other protons are random in orientation and time dependent. The z components of the local dipole fields add to or subtract from H_0 and cause a more rapid or less rapid precession than ω_0 about the z axis. In the rotating frame, these z components are responsible for the only existing precession about the z axis. Looking back at the discussion surrounding the Bloch equations in Chapter 2, the student should be able to recognize that these fields contribute to a T_2 process. We shall defer until later the discussion of T_1 processes, but it is appropriate here to identify their source. Precession of the spin *away* from the z axis is caused by transverse fields that are *static* in the rotating frame. Thus we expect local fields having components transverse to the z direction and frequency components at the Larmor frequency to contribute to T_1 processes. There is more to it than that, and we shall return to the T_1 problem later.

Let us construct a model of the longitudinal (z direction) local fields, and compute T_2. Let us presume that each spin in the sample sees a constant local field $h_L \ll H_0$ in the z direction for a time τ_c, after which it may or may not, with equal probability, reverse itself. In these terms, the problem can be phrased in terms of the famous "random walk" problem.[1] In that problem, the mean square distance traveled in the x direction, $\langle x^2 \rangle$, after n steps, each of length Λ in either the $+x$ or $-x$ direction, is

$$\langle x^2 \rangle = n\Lambda^2 \tag{3-2}$$

[1] For a derivation using arithmetical induction, see Feynman [2], vol. 1, p. 6–5.

In our problem, the unit of length is phase of precession about the z axis in the rotating frame, and the step length is $\gamma h_L \tau_c$. After time t, the number of steps n is t/τ_c; therefore, from Eq. (3-2), the mean square phase accumulated is

$$\langle \Phi(t)^2 \rangle = \frac{t}{\tau_c} (\gamma h_L \tau_c)^2 = \gamma^2 h_L{}^2 \tau_c t \tag{3-3}$$

It is perfectly within the spirit of the Bloch equations to identify T_2' as the time such that $\langle \Phi^2 \rangle = 1\ (\text{rad})^2$.[2] This criterion yields

$$\frac{1}{T_2'} = (\delta\omega)^2 \tau_c \tag{3-4}$$

where $\delta\omega = \gamma h_L$.

The particular problem we have "solved" seems a very artificial model for the behavior of internal fields in a liquid. The definitions of $\delta\omega$ and τ_c can be sharpened a great deal but at the expense of some mathematical complexity. As it stands, Eq. (3-4) provides an extremely useful estimate of T_2 in a liquid. If you wish to think of motions as being more "fluid," less jerky than the model suggests, then τ_c may be regarded as the time during which the local field changes by an amount comparable to its magnitude—a rather vague concept, to be sure, but one which might satisfy one's feeling that molecules are in continuous motion. Actually, the "jump" model has been shown by NMR techniques as well as by neutron diffraction to be a fair description of a liquid, in which a given environment around a given molecule persists for τ_c, followed by a change to another configuration, with the duration of the changing time being much less than τ_c. If one regards the constant τ_c to be an average, and the single local field to be an average over local fields caused by all possible local arrangements of nearly magnetic moments, then the picture may not look so unrealistic. And it should look quite good for describing the effect on the nuclear resonance line width of diffusion in solids.

Equation (3-4) is a reasonable estimate of T_2 as long as $\delta\omega\tau_c < 1$. If τ_c is long, the step length $\delta\omega$ is itself, when divided by γ, the width of the resonance line. So in the presence of a changing local environment, the line width is narrower than the static line width $\delta\omega$, as long as the local environment changes rapidly compared with $1/\delta\omega$.

Equation (3-4) is useful in a wider variety of circumstances than the student can presently imagine. We digress for a paragraph from the

[2] The use of the notation T_2' instead of T_2 will be explained later, when T_1 processes are discussed. The distinction between T_2 and T_2' is made here to indicate that we may not have included all possible sources of T_2 in this discussion.

main line of thought to show the application of measurements to the measurement of the diffusion constant. Let τ_c be the mean time between jumps or changes in the local environment. Imagine that a spin jumps spatially an interatomic distance a_0 each time. It proceeds, then, by random walk (in three dimensions) and travels in time t a mean square distance

$$\overline{l^2} \simeq \frac{t}{\tau_c} a_0 \tag{3-5}$$

The process described is called diffusion. It is described in continuum theory by a differential equation in the concentration c of the diffusing constituent:

$$D \nabla^2 c - \frac{\partial c}{\partial t} = 0 \tag{3-6}$$

The constant D is the diffusion coefficient. If Eq. (3-6) is solved for simple initial conditions and one-dimensional geometries, the root mean square distance the concentration spreads in time t from an initial concentration is approximately $(2Dt)^{1/2}$. Thus in Eq. (3-5) we identify

$$D \simeq \frac{{a_0}^2}{2\tau_c} \tag{3-7}$$

Since a_0 is related to the density, we have in Eq. (3-7) an expression for τ_c in terms of macroscopic quantities that can be determined by quite different, nonresonance experiments. Alternatively, we see from Eq. (3-4) that T_2 measurements can provide a value for the diffusion constant D, a number frequently hard to get if one is concerned with the self-diffusion of a molecule surrounded by identical molecules—H_2O in H_2O, for example.

To summarize Eq. (3-4) and some of the subsequent discussion, we plot log T_2 versus τ_c in Fig. 3-1. The abscissa might easily be $1/D$ or η/T, where η is the viscosity and T the absolute temperature. For justification of the last clause, see the reprint of Bloembergen's thesis [3], the original work in the field. Note in Fig. 3-1 the leveling off of T_2 to $\delta\omega^{-1}$ at a value of τ_c on the order of $\delta\omega^{-1}$, something we have justified only by a plausibility argument so far.

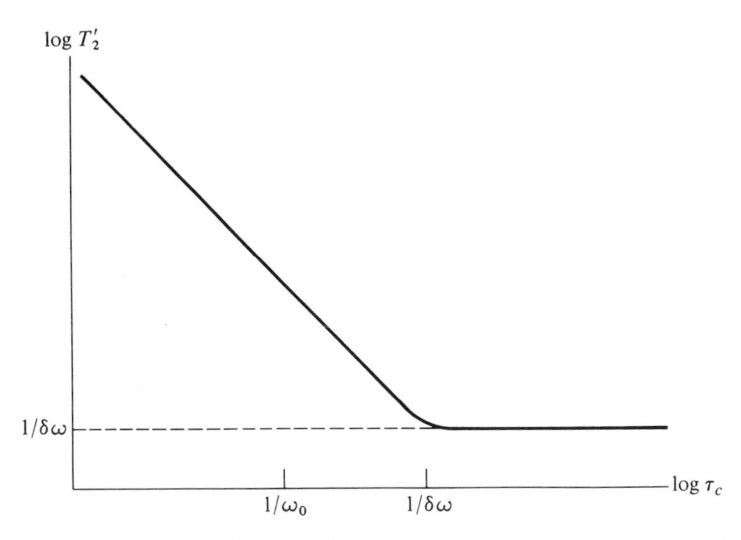

Fig. 3-1 Log T_2 versus log τ_c. Plot of Eq. (3-4) in region of its validity, $\delta\omega\tau_c < 1$.

3-3. VERY SHORT CORRELATION TIMES: $\omega_0 \tau_c \ll 1$

Suppose $\omega_0 \tau_c \ll 1$, and suppose the medium is spatially isotropic in the sense that the fluctuating internal fields point in no preferred direction. Then the magnitudes of the fields parallel to and transverse to H_0 are the same, and their frequency properties are also the same. As a result, the random walk in the transverse plane in the rotating coordinate system, produced as described in the last section, also occurs *away* from the z axis. This longitudinal relaxation, a T_1 process, is produced by transverse local fields *stationary* in the rotating frame, that is, at ω_0. The two independent and orthogonal random walks occur at *exactly the same rate* if the magnitude of the fluctuating fields at or near zero frequency is the same as at the Larmor frequency, ω_0. The simplest and most plausible description of the random internal field has that property if $\omega_0 \tau_c \ll 1$.

Since there are two orthogonal transverse components of the local field in the rotating frame, and since they act independently, the mean square angular migration of a spin *away* from the z direction is

$$\langle \phi_{T_1}(t)^2 \rangle = 2\gamma^2 h_L{}^2 \tau_c t \tag{3-8}$$

just twice as great as in Eq. (3-3). Again setting $\phi_{T_1}(T_1)^2 = 1$, we obtain

$$\frac{1}{T_1} = 2(\delta\omega)^2 \tau_c = \frac{2}{T_2'} \tag{3-9}$$

At this point, we see the reason for the introduction of the notation T_2' before Eq. (3-4). If we wish to calculate the parameter T_2, which characterizes the line width of a Lorentz line, we must include not only the inhomogeneity of the quasistatic local fields (so-called *secular* broadening), but also lifetime, or *nonsecular* broadening. Reexamination of the Bloch equation for M_x and a little thought produces the following conclusion. A T_2 process is one that causes precession of M_x away from the x axis of the rotating frame. Quasistatic fields in the z direction do this, as do fields in the y direction at ω_0. Fields in the x direction produce no effect because they exert no torque on M_x. Hence, we should have

$$\frac{1}{T_2} = \frac{1}{T_2'} + \frac{1}{2T_1} \tag{3-10}$$

where the first term is from Eq. (3-4) and the second is half of Eq. (3-9). Comparing Eqs. (3-4), (3-9), and (3-10), we see that

$$T_1 = T_2 \tag{3-11}$$

The equality holds in the limit $\omega_0 \tau_c \ll 1$ and depends, we reiterate, on spatial isotropy of the local fields. We have also filled in more of Fig. 3-1 if we regard it as a plot of T_2^{-1} or T_1^{-1} versus τ_c, rather than $(T_2')^{-1}$ versus τ_c; namely, the T_2 and T_1 curves for $\tau_c \ll 1/\omega_0$ coincide. To fill in the T_1 curve for longer τ_c we require more general analytical tools, the development of which we shall indicate in the next section.

3-4. RANDOM FREQUENCY MODULATION; SPECTRAL DENSITY

In this section we shall make heavy use of the analogy between a magnetic moment precessing at the frequency $\nu = (\gamma/2\pi)[H_0 + h_L(t)]$ and a classical, frequency modulated oscillator transmitting, for example, classical music at a center frequency of 97.8 MHz.

We shall begin by considering a proper mathematical description of frequency modulation. The simplest situation to treat is a sinusoidal frequency deviation; that is, the angular frequency of the oscillator is given by

$$\omega(t) = \omega_0 + \Delta\omega \cos qt \tag{3-12}$$

where $\Delta\omega$ is the frequency deviation and q is the frequency of modulation. We now need an expression for $A(t)$, the amplitude of the fm oscillator as a function of time. One approach might be to write

$$A(t) = A_0[\cos \omega(t)t]$$

We shall not number that equation, since it is wrong, just as it is wrong to write that the distance a car travels in t to be $d = vt$ if v is not a constant. The argument of a trigonometric function is a *phase*, and the phase accumulated between $t' = 0$ and t with frequency given by Eq. (3-12) is

$$\phi(t) = \int_0^t \omega(t')\,dt' = \omega_0 t + \frac{\Delta\omega}{q} \sin qt \tag{3-13}$$

Now we can safely write, complete with equation number,

$$A(t) = A_0 \cos\left[\omega_0 t + \frac{\Delta\omega}{q} \sin qt\right] \tag{3-14}$$

The ratio $\Delta\omega/q$ is an extremely important parameter for future discussions:

$$m = \frac{\Delta\omega}{q} \tag{3-15}$$

where m is called the *modulation index*. Its size relative to unity will be crucial in many applications.

We wish to know the frequency spectrum of Eq. (3-14). First, it is more convenient to rewrite Eq. (3-14) using the trigonometric identity $\cos(a + b) = \cos a \cos b - \sin a \sin b$:

$$A(t) = A_0[\cos \omega_0 t \cos(m \sin qt) - \sin \omega_0 t \sin(m \sin qt)] \tag{3-16}$$

From immediate inspection we can see that the main frequency components are at ω_0, and the other spectral content is caused by the terms $\cos(m \sin qt)$ and $\sin(m \sin qt)$. They are periodic with period $2\pi/q$; that is, the argument of these terms repeats itself with that period. That fact makes them prime candidates for expansion in Fourier series. Consider the cosine function

$$\cos(m \sin qt) = \sum_{n=0}^{\infty} a_n(m) \cos nqt \tag{3-17}$$

We can use only cosine terms in the expansion since the cosine is an even function and the coefficients of sine terms would necessarily vanish. The coefficients $a_n(m)$ are found by multiplying both sides of Eq. (3-17) by $\cos n'qt$ and averaging over a period T. If you do so, you find the following integral expression:

$$a_{n'}(m) = \frac{1}{T} \int_0^T \cos n'qt \cos(m \sin qt)$$

This expression may be developed in a power series in $m \sin qt$, which becomes, upon integration, a power series in m. Fortunately, the labor has already been done; consultation of a complete set of mathematical tables [4] under "Bessel Functions" shows the coefficients of Eq. (3-17) to be Bessel functions of integral order:

$$\cos(m \sin qt) = J_0(m) + 2 \sum_{k=1}^{\infty} J_{2k}(m) \cos 2kqt \tag{3-18a}$$

and

$$\sin(m \sin qt) = 2 \sum_{k=0}^{\infty} J_{2k+1}(m) \sin[(2k+1)qt] \tag{3-18b}$$

The functions $J_n(m)$ are Bessel functions of integral order of the first kind. Although they are surely less familiar than trigonometric functions, the student should not be put off by them. A graph of the first three is shown in Fig. 3-2. Note $J_0(0) = 1$, and $J_n(0) = 0$, $n = 0$. Also note that $J_0(m)$ deviates as m^2 from its value at $m = 0$ for small m, and the others begin from zero as m^n. A qualitative look at Fig. 3-2 and these remarks are all we require of the Bessel functions. The expressions (3-18) must be put back into (3-16):

$$A(t) = J_0(m) \cos \omega_0 t + \sum_{n=-\infty}^{\infty} J_{|n|} \cos(\omega_0 + nq)t(\text{sgn } n) \tag{3-19}$$

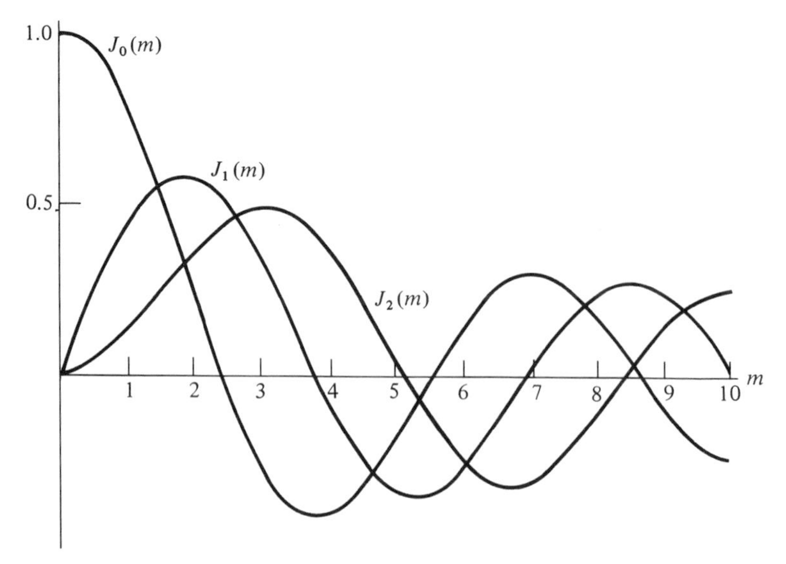

Fig. 3-2 Bessel functions of the first kind, $J_n(m)$, versus m, for $n = 0, 1, 2$.

The notation (sgn n) means to multiply by $(+1)$ when $n > 0$ and by (-1) when $n < 0$. Again, Eq. (3-19) is most valuable to use in graphical form, as in Fig. 3-3, which gives the frequency spectrum of the power, proportional to $A(t)^2$. The graph has been presented with $\Delta\omega =$ constant, because in our applications, the local field, which causes $\Delta\omega$, is a fixed characteristic of the substance, whereas q may often be changed within a given sample, by changing the temperature, for example. (We shall present one situation, however, in which it is q that is fixed by the nature of the substance and $\Delta\omega$ that is changed by the experimenter.) The thing to notice about Fig. 3-3 is that the power spectrum is mostly contained within $\Delta\omega$ of ω_0, and when $m < 1$, which means $q > \Delta\omega$, the power is mainly in the center frequency ω_0. The sidebands are still spaced by the modulation frequency q but are small in amplitude. If you try to change the frequency back and forth too rapidly, the major effect is not to change it at all.

Although there is only a qualitative resemblance between the problem of sinusoidal frequency modulation and our problem of motional narrowing, the results of the preceding analysis are already suggestive. If we make an arbitrary connection between the spectrum of Fig. 3-3 and the line width in a nuclear resonance, we can see the origins of the broadening that takes place as τ_c increases and becomes on the order of $1/\delta\omega$ in Fig. 3-1. For all values of τ_c such that $\delta\omega\tau_c < 1$, the "modulation index" is less than 1; that is, $\delta\omega\tau_c$ is roughly the same as the modulation index.

There is, of course, a vast difference between the case of sinusoidal modulation and the random fluctuations of the local field in a liquid. To

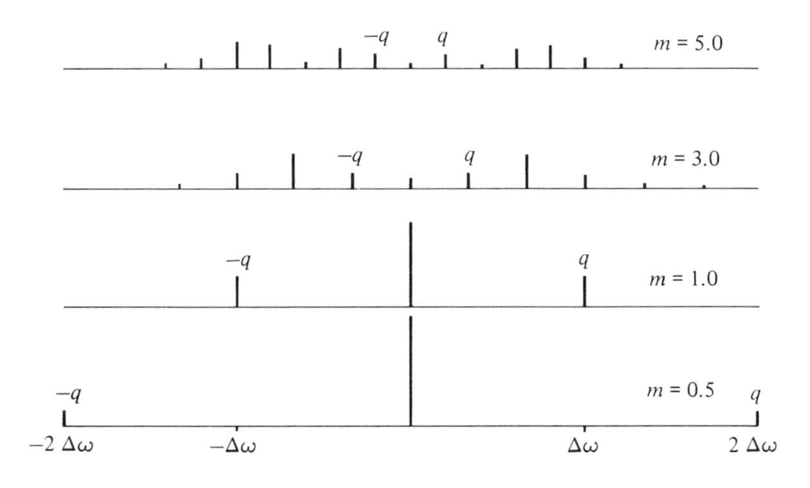

Fig. 3-3 Fourier components of the power spectrum of Eq. (3-19) for $m = 0.5$, 1.0, 3.0, 5.0.

sharpen the distinction, we make the following analysis. Let $h_L(t)$ be the local field seen by a nucleus at some arbitrary time t. Form the product $h_L(t)h_L(t + \tau)$ and average over all time for a particular nucleus. That process defines $f(\tau)$:

$$f(\tau) = \langle h_L(t)h_L(t + \tau)\rangle \tag{3-20}$$

where the brackets $\langle\ \rangle$ indicate average over time t and $f(\tau)$ is the *autocorrelation function* of h_L. It is independent of t since the sample is assumed to be homogeneous and in thermal equilibrium; there is nothing special about any particular time. Consider what we expect of $f(\tau)$ as τ becomes large. If the sample is large, if the local field is produced by many nuclei undergoing random motions, then we ought reasonably to expect no relation between $h_L(t)$ and $h_L(t + \tau)$. Since h_L can be positive or negative, and since, in fact, the temporal randomness of h_L means $\langle h_L(t)\rangle = 0$, we expect

$$\lim_{\tau\to\infty} f(\tau) = 0 \tag{3-21}$$

to be reasonable behavior for $f(\tau)$ at large τ. Contrast that behavior with the autocorrelation function of Eq. (3-12):

$$\begin{aligned}\langle\omega(t)\omega(t + \tau)\rangle &= \langle(\omega_0 + \Delta\omega\cos qt)[\omega_0 + \Delta\omega\cos q(t + \tau)]\rangle \\ &= \omega_0{}^2 + \Delta\omega^2\left(\frac{q}{2\pi}\int_0^{2\pi/q}\cos qt\cos q(t + \tau)\,dt\right) \\ &= \omega_0{}^2 + \Delta\omega^2\,\frac{\cos q\tau}{2}\end{aligned} \tag{3-22}$$

The leading term, $\omega_0{}^2$, is as expected, but the second term certainly does not vanish. Therefore, it is clear that a simple modulation such as Eq. (3-12) fails to satisfy one's intuitive requirements for random modulation.

The simplest form of $f(\tau)$ that satisfies all the requirements is an exponential:

$$f(\tau) = \langle h_L(t)^2\rangle \exp -|\tau|/\tau_c \tag{3-23}$$

Equation (3-23) reintroduces the correlation time τ_c. The average over time of the square of the local field, $\langle h_L(t)^2\rangle$, is the same as the average of $h_L{}^2$ over all the nuclei in the sample by a fundamental hypothesis of statistical mechanics [5]. We see that we are able to specify the local field, and hence the random frequency modulation of each "nuclear oscillator" by that local field, less precisely than we did in the simple case of sinusoidal modulation, but we do preserve some points of similarity. These include the correspondence between $(\gamma h_L)^2$ and $\Delta\omega^2$, and the similarity between the parameter q of Eq. (3-22) and $1/\tau_c$ of Eq. (3-23).

At this point, our resort to plausibility arguments must come to an end, because the subsequent development relies on theory that uses the density matrix of quantum statistical mechanics and time dependent perturbation theory. The full theory is nothing less than a derivation of the Bloch equations from the basic principles of quantum statistical mechanics. Our previous paragraphs have attempted to establish a climate of acceptance in the student's mind for the results. The central formula of the theory is the Fourier transform of the correlation function $f(\tau)$, which is $j(\omega)$, the spectral density function:

$$j(\omega) = \tfrac{1}{2}\int_{-\infty}^{\infty} \langle h_L(t)h_L(t+\tau)\rangle e^{-i\omega\tau}\, d\tau \tag{3-24}$$

Equation (3-24) is one of a pair of integrals known as the Wiener–Khintchine relations (see reference [5], p. 586). In the case of the particular form of $f(\tau)$ given by Eq. (3-24),

$$j(\omega) = \frac{\langle h_L{}^2\rangle \tau_c}{1 + \omega^2 \tau_c{}^2} \tag{3-25}$$

The spectral density function determines the effectiveness of the local field in producing transitions between quantum mechanical spin states. The relaxation rates $1/T_1$ and $1/T_2$ are determined by $j(\omega)$. To write down the final results, we must specify $j(\omega)$ more completely. Since the local field has the usual x, y, and z components, a more general $j(\omega)$ may be written

$$j_{\alpha\beta}(\omega) = \tfrac{1}{2}\int_{-\infty}^{\infty} h_\alpha(t)h_\beta(t+\tau)e^{-i\omega\tau}\, d\tau \tag{3-26}$$

where $\alpha, \beta = x, y, z$. If the local field is truly random, then $\langle h_\alpha(t)h_\beta(t+\tau)\rangle = 0$, for $\alpha \neq \beta$, and the only components of $j_{\alpha\beta}$ are $j_{xx}(\omega)$, $j_{yy}(\omega)$, and $j_{zz}(\omega)$. In terms of the spectral density function, the complete equation for T_2^{-1} that replaces Eqs. (3-4) and (3-10) is

$$T_2^{-1} = \gamma^2[j_{zz}(0) + j_{yy}(\omega_0)] \tag{3-27}$$

The first term is the same as Eq. (3-4), which we obtained by the random walk argument, and the second is the "lifetime broadening" effect we discussed prior to Eq. (3-10). In terms of h and τ_c, from Eq. (3-25) we get

$$T_2^{-1} = \gamma^2\left(\langle h_z{}^2\rangle\tau_c + \langle h_y{}^2\rangle \frac{\tau_c}{1 + \omega_0{}^2\tau_c{}^2}\right) \tag{3-28}$$

The assumption of spatial isotropy, made to obtain Eq. (3-10), is that $\langle h_x{}^2\rangle = \langle h_y{}^2\rangle = \langle h_z{}^2\rangle \equiv h_L{}^2$. The result for T_1^{-1} is

$$T_1^{-1} = \gamma^2[j_{xx}(\omega_0) + j_{yy}(\omega_0)] = \frac{2\gamma^2 h_L{}^2\tau_c}{1 + \omega_0{}^2\tau_c{}^2} \tag{3-29}$$

We can now complete Fig. 3-1 by including the portion of the T_1 curve for $\tau_c > 1/\omega_0$. The full curve is shown in Fig. 3-4. At $\tau_c = 1/\omega_0$, T_1 goes through a minimum, then increases, whereas T_2 continues to decrease. Although these features are an obvious analytical consequence of

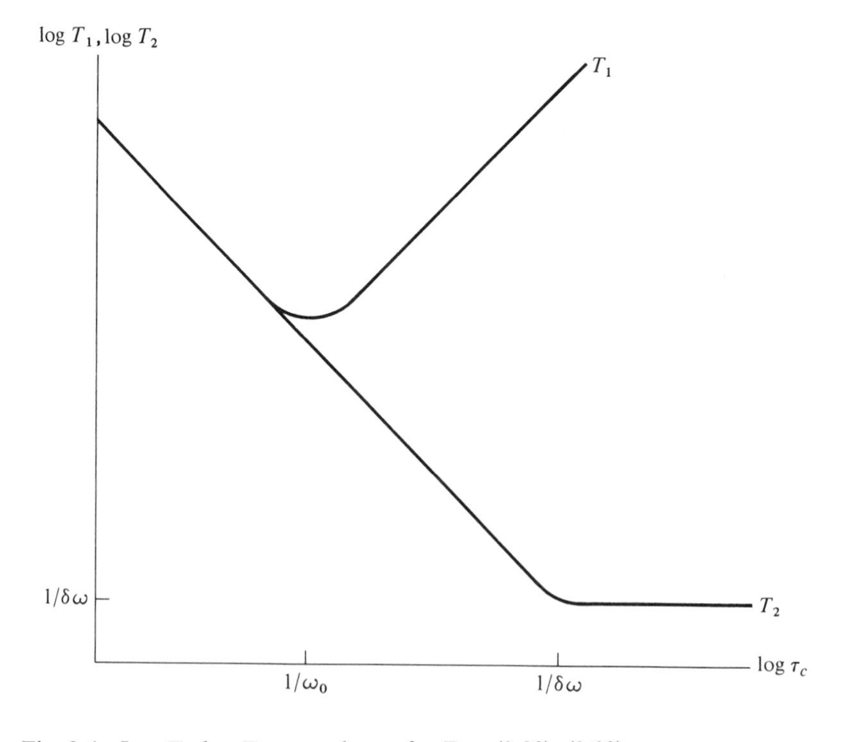

Fig. 3-4 Log T_1, log T_2 versus log τ_c for Eqs. (3-28), (3-29).

Eqs. (3-28) and (3-29), it is instructive to see how they can be remembered easily by examining one further property of $j(\omega)$, Eq. (3-25):

$$\frac{1}{h_L{}^2}\int_0^\infty j(\omega)\,d\omega = \int_0^\infty \frac{\tau_c d\omega}{1+\omega^2\tau_c{}^2} = \int_0^\infty \frac{dx}{1+x^2} = \frac{\pi}{2} \tag{3-30}$$

The area under $j(\omega)$ versus ω is a constant, independent of τ_c. Several $j(\omega)$ curves with this property are plotted in Fig. 3-5. When τ_c is short [curve (1) of Fig. 3-5], $j(\omega_0) = j(0)$, and $T_1 = T_2$. $j(\omega_0)$ is largest for curve (2)—hence the maximum in the relaxation rate there, or the T_1 minimum of Fig. 3-4. For long τ_c, $j(\omega_0)$ decreases again as the Larmor frequency falls far out in the tail of $j(\omega)$, accounting for the rise in T_1 for longer τ_c.

Nothing in Eq. (3-28) explains the constancy of T_2 when τ_c is longer than $1/\gamma h_L$. At this point, the theory breaks down, but the reason for it and the result can be seen in Fig. 3-3. The spectrum for $m \gg 1$ covers only the range $\Delta\omega$ as the modulation index increases. The local field is essentially static. Under these conditions, the Bloch equations are not valid, the line shape of the resonance, $\chi''(\omega)$, is not Lorentzian. We shall leave this case to the next chapter.

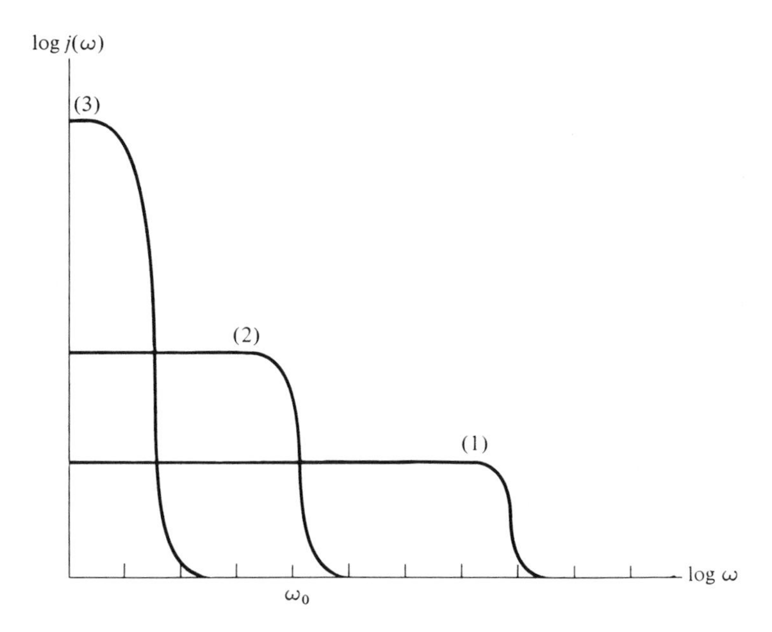

Fig. 3-5 (1) $j(\omega)$ versus ω, τ_c less than ω_0. (2) τ_c equal to ω_0. (3) τ_c greater than ω_0.

3-5. SOME APPLICATIONS

Qualitatively, the effect of rapid nuclear motion is to average out internal fields, or so the previous parts of this chapter would lead us to believe. In fact, we must be very careful to make a distinction between internal fields that can be averaged out and those that cannot. Equation (3-1) gives the expression for the field at a vector distance **r** from a point dipole

$$\mathbf{H}_d = -\frac{\boldsymbol{\mu}}{r^3} + 3\frac{\boldsymbol{\mu}\cdot\mathbf{r}}{r^5}\mathbf{r} \tag{3-1}$$

Consider, for example, the z component of the field produced at the center of a unit sphere by a dipole on the surface. If the dipole points in the z direction, then

$$H_z = -\mu(1 - 3\cos^2\theta)r^{-3}$$

where θ is the polar angle in the conventional spherical coordinate labeling. If the dipole is allowed to roam over the surface of the sphere, the average field at the center is

$$\frac{1}{4\pi}\int_{\phi=0}^{2\pi} d\phi \int_{\theta=0}^{\pi} (1 - 3\cos^2\theta)\sin\theta\, d\theta = 0$$

since $\langle\cos^2\theta\rangle = \frac{1}{3}$ when averaged over the solid angle. The demonstration of the equivalent result for other components of $\mathbf{H}_d$ and arbitrary orientation of $\boldsymbol{\mu}$ is tedious, and, in fact, unnecessary, but the result still holds.

Another type of *local* field that can be reduced by rapid motion is the inhomogeneity of the magnet. In this case, the inhomogeniety of H_z is all that counts. Suppose, as is likely to be the case in the typical electromagnet, the inhomogeneity has cylindrical symmetry, and suppose the maximum deviation from the average H_0 is ΔH. If a given nucleus can be forced to sample the range of fields over the sample volume rapidly enough, the total effect of the local field will be reduced according to Eq. (3-4). The nuclei may be caused to sample the magnet's range of fields by spinning the sample about an axis perpendicular to the field direction. A typical geometry is shown in Fig. 3-6. To achieve narrowing, ω must be so large that the modulation index, $m = (\gamma\,\Delta H/\omega)$, is less than one. If each spin samples essentially the same magnetic fields during each revolution, the frequency modulation produced by the spinning is periodic, although probably not sinusoidal. Our original analysis of frequency modulation is useful, though, and we see that the criterion $m < 1$ is sufficient to produce *spinning sidebands* which are farther away

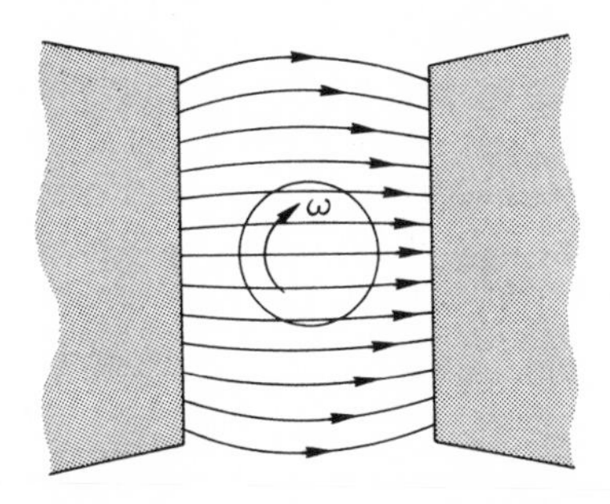

Fig. 3-6 Spherical sample in an inhomogeneous field. Sample is spun at angular frequency ω about an axis perpendicular to the page.

than $\gamma\,\Delta H$. For a typical field inhomogeneity of 10^{-4} G, such as is found in electromagnets made for chemistry applications, a spinning frequency $(\omega/2\pi)$ of a few cycles per second is sufficient. If turbulence within the sample causes a given spin to sample the available fields in a more random fashion, we require the width produced by the field inhomogeneity after narrowing to be comparable to, or less than, the natural width. The requirement from Eq. (3-9) may be expressed

$$\frac{1}{T_2} < \frac{1}{T_2^*} = \frac{(\gamma\,\Delta H)^2}{\omega} \tag{3-31}$$

For typical hydrogenous liquids, where $T_2 \simeq 5$ sec, and with a magnet inhomogeneity of 10^{-4} G, ω again must be a few cycles per second. That spinning rate is easily achieved, and commercial apparatus is routinely supplied with sample spinning attachments.

Applications to chemistry

The internal magnetic fields of purely dipolar origin are not usually very interesting because they are well understood. There are other magnetic fields of intermolecular or atomic origin that are almost always smaller than the static dipolar fields but that are not almost averaged to zero by motional narrowing. One such field goes by the name *chemical shift*. The actual field seen by the nucleus is the external field H_0 plus fields proportional to H_0 created by electronic currents within the nucleus's own atom or molecule. The resonance field is thus

$$H_R = H_0(1 - \sigma) \tag{3-32}$$

where σ, the chemical shift, may be positive (diamagnetic) or negative (paramagnetic). The origin of σ is, roughly speaking, the same as the origin of atomic diamagnetism, and it depends in detail on the electronic wavefunctions of the ground and excited states of the molecule—hence its interest to theoretical chemistry. Those interested in more discussion of the physics of the chemical shift should consult one of the standard chemistry texts on the subject (see references at end of this chapter) or Slichter [1].

In magnitude, $|\sigma|$ ranges from zero to several hundred parts per million. Most chemical applications of the existence of the shift are concerned not so much with its absolute magnitude but with its magnitude and sign relative to an arbitrary standard that is defined to have zero shift. The existence of a vast quantity of empirical information on chemical shifts of protons and other nuclei common in organic molecules (notably ^{13}C, ^{19}F, ^{31}P) has proved to be extremely useful in identifying the position on a complex molecule of a nucleus responsible for a particular resonance line. Figure 3-7 reproduces the famous resonance of ethyl alcohol (taken with very low resolution by present standards) in which the different chemical shifts associated with protons on the various groups in the molecule are clearly distinguishable.

One exception to the usual unimportance of absolute shifts as compared to relative shifts occurs in the evaluation to high precision of the atomic constants. The magnetic moment of the proton is one quantity it is necessary to know with great precision because of its involvement with the hyperfine structure (hfs) of hydrogen, and the hfs of hydrogen has an

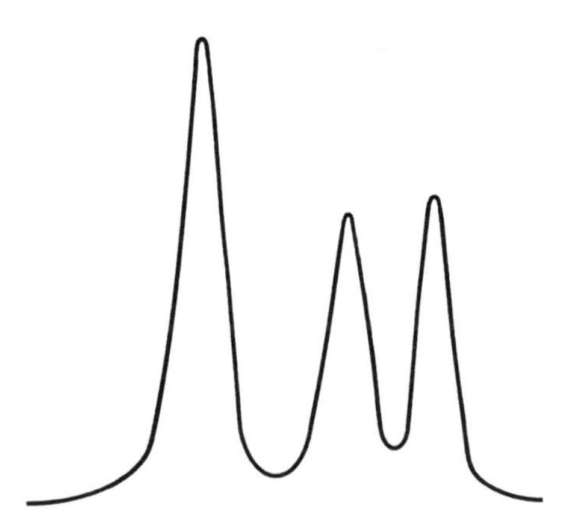

Fig. 3-7 Relatively low resolution spectrum of ethyl alcohol, showing the relative chemical shifts of CH_3, CH_2, and OH protons [after J. T. Arnold, S. S. Dharmatti, and M. Packard, *J. Chem. Phys.* **19**, 507 (1951)].

important bearing on the value of the fine structure constant α. For such purposes, the actual magnitude of the shielding of the protons by the electrons on H_2 and H_2O is of great interest. There is no experimental answer to the question; the chemical shift must be calculated from molecular wavefunctions.

There is a small but important interaction between nuclei in molecules that would be totally obscured if dipole-dipole interactions were not nearly averaged away. The interaction can be expressed in the form of an effective of *spin Hamiltonian* of the form

$$\mathscr{H} = \hbar A \mathbf{I}_1 \cdot \mathbf{I}_2 \tag{3-33}$$

The constant A is small; A's as small as a fraction of a cycle per second have been measured. The coupling of this form between H and D in the HD molecule, about 40 Hz, is relatively large as such couplings go. The coupling Eq. (3-33) is variously known as the scalar spin-spin interaction or scalar exchange interaction. It depends, clearly, on the relative orientation of the two spins involved, not on their spatial positions. Hence, it is invariant under molecular tumbling since the nuclear spin directions are almost totally decoupled from the spatial orientation of the molecule.

It is not the purpose of this book to discuss in any detail the origins of such interactions as Eq. (3-33), interesting as they may be. But a qualitative explanation is not out of place. For definiteness, focus on the HD molecule in a large magnetic field. The electronic wavefunctions for the molecule with the proton in the $m_I = +\frac{1}{2}$ state differ very slightly from those with the proton in the $m_I = -\frac{1}{2}$ state because of the magnetic interactions of the electrons with the proton. The magnetic interaction of the deuteron with the electron cloud thus depends on the magnetic quantum state of the proton. The same description could be made starting with the deuteron; this state of affairs is described by Eq. (3-33).

What effect does Eq. (3-33) have on the resonance spectrum? The total interaction Hamiltonian of the nuclear spins with each other and with the external field (excluding the dipole-dipole term) is

$$\mathscr{H} = (-\gamma_1 \hbar \mathbf{I}_1 - \gamma_2 \hbar \mathbf{I}_2) \cdot \mathbf{H}_0 + \hbar A \mathbf{I}_1 \cdot \mathbf{I}_2 \tag{3-34}$$

If $\mathbf{H}_0$ is in the z direction, and if $A \ll \gamma H_0$, then Eq. (3-34) is well approximated by

$$\mathscr{H} \simeq (-\gamma I_{z1} - \gamma I_{z2}) \hbar H_0 + \hbar A I_{z1} I_{z2} \tag{3-35}$$

Equation (3-35) is an approximation, of course, since it ignores the terms $I_{1x}I_{2x} + I_{1y}I_{2y}$ in the scalar product $\mathbf{I}_1 \cdot \mathbf{I}_2$; it includes only the diagonal elements of Eq. (3-34). It is also seen to be plausible from the old-fashioned vector model by using the same type of argument as is used in the vector model derivation of the Landé g factor. The inequality that must be obeyed for Eq. (3-35) to be a good approximation to Eq. (3-34) is that $(\gamma_H - \gamma_D)H_0 \gg A$. More will be said about this approximation when we discuss the exchange interaction between electronic spins.

The resonant frequencies for nuclei of type 1 are

$$\omega = \gamma_1 H_0 + Am_2 \tag{3-36}$$

There are thus $2I_2 + 1$ lines centered at $\gamma_1 H_0$. Similarly, for nucleus I_2 there are $2I_1 + 1$ lines centered at $\gamma_2 H_0$. The schematic appearance of the resonances for HD are shown in Fig. 3-8. We would be wrong to expect an analogous structure for H_2 or D_2. Although there is a coupling of type (3-33) between the two identical and *chemically equivalent* protons or deuterons, there is no experimental manifestation of the coupling in a magnetic resonance experiment. The absence of a splitting is a quantum mechanical effect; it is caused by both the identity of the moments and the

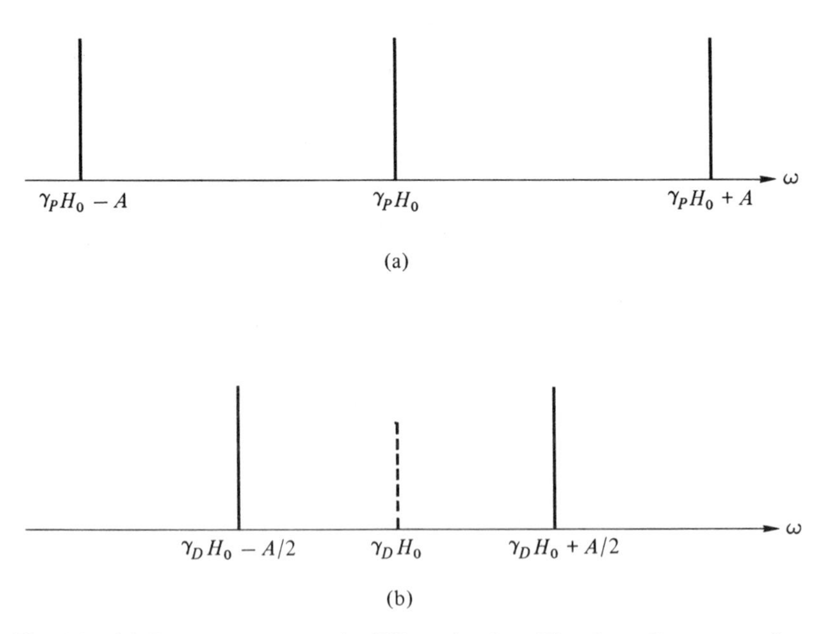

Fig. 3-8 (a) Proton resonance in HD molecule. The three lines come from the $2I_D + 1 = 3$ orientations of the deuterium nucleus. (b) Deuterium resonance in HD. The splitting, $A = 2\pi \cdot 43$ Hz, is independent of the field in which the resonance is done, and is the same for D as for H.

scalar nature of the coupling. An elegant proof by Slichter of this fact is to be found in the paper that gave the first exposition of the experimental consequences of the scalar coupling [6].

Suppose that for some reason the deuteron in HD relaxed much more rapidly than the proton (as it conceivably might because it has a quadrupole moment). The effective internal field as seen by the proton can be written

$$H_{\text{int}}\,(\text{at proton}) = \frac{A}{\gamma_p}\, m_{\text{D}} \tag{3-37}$$

and the associated $\delta\omega$ [in the sense, now, of our random walk arguments which led to Eq. (3-4)] is A. If the deuteron relaxes at the rate $(1/T_{1\text{D}})$, the effective field at the proton changes at that rate also. The process is characterized by a modulation index $m \simeq AT_{1\text{D}}$. The structure on the proton resonance of Fig. 3-7 is preserved if $m > 1$ and is washed out when $m < 1$.

It is sometimes useful to simulate artificially the process previously described to aid in the unraveling of complex spectra. We use again the example of HD, and revoke the assumption of rapid relaxation of the deuterium. If we apply a large transverse rf field $H_{1\text{D}}$ at the deuterium Larmor frequency, the deuterium spin precesses around it at $\gamma_{\text{D}} H_{1\text{D}}$. The proton, from Eq. (3-37), sees a sinusoidally varying m_{D} (using loose, classical language) at $\gamma_{\text{D}} H_{1\text{D}} = q$, in the notation of Section 3-4. Again, the modulation index is

$$m = \frac{\Delta\omega}{q} = \frac{A}{\gamma_{\text{D}} H_{1\text{D}}}$$

and the structure on Fig. 3-8a is washed out when $m \ll 1$ or $\gamma_{\text{D}} H_{1\text{D}} \gg A$. You should note, incidentally, that the separation of the lines in Fig. 3-8b is, on a magnetic field scale, $(A/\gamma_{\text{D}}) \ll H_{1\text{D}}$, so that one can ignore the fine structure of the line in describing the motion of the deuterium nucleus in the presence of $H_{1\text{D}}$. The deuterium resonance will, of course, be heavily saturated. There may be other effects on the proton resonance in addition to the collapse of the triplet in Fig. 3-8a. I refer particularly to a change in intensity of the proton resonance because of the "nuclear Overhauser effect." A brief discussion of the Overhauser effect is given in Chapter 5. A very recent technique reminiscent of the previous discussion is presently still being developed, in which the effect of nuclear motion in narrowing dipolar line widths is simulated in solids by application of a judiciously chosen sequence of intense rf pulses. The problem is much more complicated than the analogous "decoupling" of the H and D spins (as the preceding process is known in chemistry), but the general idea is much the same [7].

A paramagnetic resonance example

The consequences of the scalar spin-spin interaction are to be found in other resonance experiments. The scalar interaction really differs from the dipolar interaction more than it may seem to. As far as resonance experiments are concerned, it was discussed originally in reference to paramagnetic resonance line shapes in solids in the late 1940's by Van Vleck [8], and then, in the context of the present discussion, by Slichter [6], and, in a different context, by Anderson [9]. Although the problem arose in paramagnetic resonance, which we are not going to discuss in this volume, if a couple of statements are taken on faith, the discussion fits well into the present chapter.

As an example we shall discuss the paramagnetic salt $CuSO_4 \cdot 5H_2O$. All one need know about it is that copper sulfate pentahydrate exhibits a paramagnetic resonance with roughly the free electron gyromatic ratio, so a typical observation of the resonance is made at microwave frequencies in fields of a few thousand gauss. The word "roughly" was chosen carefully, for, in fact, the g value of the spin, which is on the Cu^{2+} ion, is neither 2.00, as with a free electron, nor is it independent of the direction of the external field relative to the symmetry axes of the crystal. That is, g is not a scalar but a tensor. In addition, there are two types of Cu^{2+} ions in each unit cell[3] of the crystal. They differ from each other only in the orientation, relative to the crystal axes, of the g tensor. At some orientations of the field relative to the crystal, the resonance looks as in Fig. 3-9a. In Fig. 3-9a, the field is around 10^4 G; hence, the microwave frequency is around 30 GHz and the wavelength λ is about 1 cm. The separations of the peaks is typically 1 kG. There is one other characteristic of copper sulfate pentahydrate: an exchange interaction between the Cu^{2+} spins

$$\mathscr{H}_{\text{exch}} = \hbar J \mathbf{S}_1 \cdot \mathbf{S}_2 \tag{3-38}$$

In this system, $S_1 = S_2 = \frac{1}{2}$, and J is the "exchange constant." [It is the same as A in Eq. (3-33), as far as we are concerned, but the physics of its origin is rather different.] Now, in contrast to the HD example, different things happen. For HD, the inequality $\gamma H_0 \gg A$ was satisfied by about six orders of magnitude, and, in addition, the inequality

[3] A unit cell is an elementary arrangement of one or more chemical formula units that may be arranged in a repeated fashion to make up the macroscopic crystal.

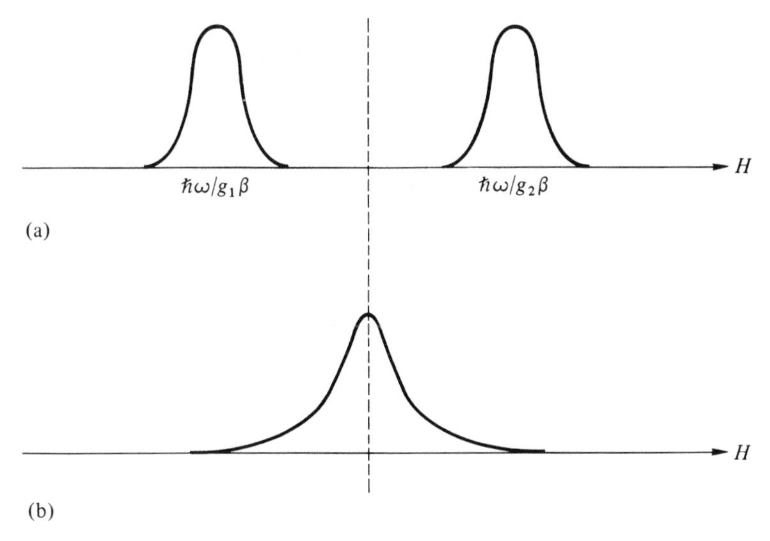

Fig. 3-9 Paramagnetic resonance absorption spectrum of two-spin system with exchange interaction $\hbar J$ between the spins. (a) $|g_2 - g_1|\beta H_0 > \hbar J$. (b) $|g_2 - g_1|\beta H_0 \ll \hbar J$. ω is the (constant) microwave frequency.

$(\gamma_P - \gamma_D)H_0 \gg A$ is satisfied by about the same amount. The J in Eq. (3-38) is, in fact, large enough so that at lower fields and microwave frequencies one may have $\hbar J > (g_1 - g_2)\beta H_0$. Figure 3-9b shows the appearance of the microwave spectrum at 10 GHz and about 3000 G, where $(g_1 - g_2)H_0\beta/\hbar$ is about 10^9 Hz.

What is the explanation of this phenomenon? Classically, one uses the following language. The constant J is taken to be the rate at which the resonant spin on sites 1 and 2 "exchange," so that the electron spins precess at frequency $g_1\beta H_0/\hbar$ for a while and then at $g_2\beta H_0/\hbar$. If the rate at which the exchange between these two Larmor frequencies is faster than $(g_2 - g_1)\beta H_0/\hbar$, then, in fm language, the modulation index is less than one,

$$m = \frac{|g_2 - g_1|\beta H_0}{\hbar J} < 1$$

and the line is "exchange narrowed." A comparison of Fig. 3-9a and 3-9b suggests that $J \sim 5 \times 10^9$ Hz, a result confirmed by other types of measurements.

The classical interpretation we have somewhat artificially forced on the discussion really does have some legitimate basis. In going from Eq. (3-34) to Eq. (3-35), we ignored the terms (using present notation)

$$S_{1x}S_{2x} + S_{1y}S_{2y} = \tfrac{1}{2}(S_{1-}S_{2+} + S_{1+}S_{2-}) \tag{3-39}$$

where $S_{\pm}$ are spin operators that have matrix elements only between spin states differing by $m_s \pm 1$. The states of the two-spin Hamiltonian may be labeled by (m_1, m_2). The operator $S_{1-}S_{2+}$ has as its only nonvanishing matrix element the quantity $\langle -\frac{1}{2}, \frac{1}{2}|S_{1-}S_{2+}|\frac{1}{2}, -\frac{1}{2}\rangle$ in the two-spin system, $S_1 = S_2 = \frac{1}{2}$. The off diagonal terms [in the (m_1, m_2) representation] cause mutual spin flips between the spins—in a sense, exchanging their spin states. But the proper quantum mechanical view of the modulation index m is that it represents the ratio of the energy difference between the states connected by $\hbar J\mathbf{S}_1 \cdot \mathbf{S}_2$ and the magnitude of the off diagonal elements, which is roughly $\hbar J$. If that ratio is much greater than one, then the diagonal terms alone provide a good description of the energy levels. In other words, the (m_1, m_2) representation is a good one. If the off diagonal terms are larger than the difference between diagonal ones—that is, if $\hbar J \gtrsim E(m_1 + 1, m_2 - 1) - E(m_1, m_2)$—then the (m_1, m_2) description is not a good one. The qualitative solution to the problem can be guessed, though, by resorting to the analogy with the fm oscillator. The whole problem is exactly the same one, in fact, as is faced in passing from the weak to the strong field case in atomic spectra (Zeeman effect to Paschen–Back effect).

Breadth of spectral lines: pressure broadening and pressure narrowing

The visible and ultraviolet spectral lines emitted from a mercury discharge tube are rather broad—usually much broader than the natural width that one would expect from the level lifetimes of the isolated atoms. The two most common causes of the excess width are pressure broadening and the Doppler effect. Consider pressure broadening first.

When an atom in an excited electronic state encounters another atom—say, for definiteness, one of the same kind in the ground state—the collision is almost always a strong one. It is strong in the sense that during the time of collision the system is perhaps better described as a combined one, with the electronic wavefunctions, and consequently their energies, quite different from the isolated atom. The collision may stimulate a transition to the ground state, in which case the line width is lifetime limited and is determined by the mean time between collisions and the probability of a transition to the ground state. The other possibility is that after the collision there is one atom in the excited state and one in the ground state. But the quantum mechanical phase difference between the excited states before and after collision is determined by the exact details of the collision —details that vary uncontrollably from collision to collision. The classical description, first proposed by Lorentz long before quantum mechanics, is of a radiating oscillator having a phase occasionally interrupted and restarted at an arbitrary phase. Phase interruption gives a line width of the radiated spectrum proportional to the reciprocal of the mean time between interruptions. The procedure is to be contrasted with motional narrowing, a process in which the frequency of the oscillator is changed rather than its phase. It is, in fact, the essence of motional narrowing that the continuity of phase is preserved. The processes are contrasted in Fig. 3-10.

The Doppler effect is important to the breadth of spectral lines, on the other hand, when the radiating gas is rare enough that an excited atom may continue so long on its course that the observer receives a Doppler-shifted frequency from the atom as a result of its relative velocity. If the radiating atom—think of it as a classical oscillator again—is moving relative to the observer with the component of velocity along the line of sight v_z, the received frequency is

$$\nu = \nu_0\left(1 \pm \frac{v_z}{c}\right) \tag{3-40}$$

where ν_0 is the radiated frequency in the rest frame of the atom and c is the velocity of light. [There are higher order terms, but $(v_z/c) < 10^{-4}$ for cases of interest, so we may neglect them.] The velocity distribution is given by the Maxwellian distribution

$$P(v_z)\,dv_z = \left(\frac{M}{2\pi kT}\right)^{1/2} \exp\frac{-Mv_z^{\,2}}{2kT}\,dv_z \tag{3-41}$$

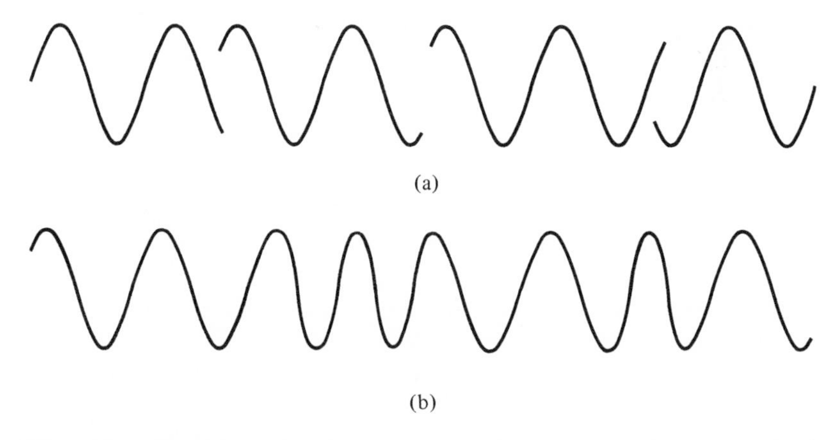

Fig. 3-10 Schematic illustration of the distinction between (a) random phase interruption and (b) random frequency modulation. In (b), phase continuity is preserved.

Equation (3-41) gives the probability that an atom of mass M in a gas at temperature T has the z component of its velocity between v_z and $v_z + dv_z$. From Eq. (3-40), v_z is related to $\nu - \nu_0$, so Eq. (3-41) may be written

$$P(v)\,dv = \left(\frac{M}{2\pi kT}\right)^{1/2} \exp\frac{-Mc^2(\nu-\nu_0)^2}{2kT{\nu_0}^2}\,dv \tag{3-42}$$

Equation (3-42) describes a Gaussian line shape with fractional width

$$\frac{\Delta\nu}{\nu_0} = 2\left(\frac{2kT}{Mc^2}\right)^{1/2} \tag{3-43}$$

Equation (3-43) is in easy form for quick estimates, since $Mc^2 \simeq 1$ GeV for hydrogen, and $kT = (1/40)$ eV for $T = 300°$K. The fractional width, $\Delta\nu/\nu_0$, is usually about 10^{-6} from lifetime broadening, at room T, and for hydrogen, Eq. (3-43) is about 1.5×10^{-5}. For heavy atoms ($M \gtrsim 100$), the Doppler width is an order of magnitude smaller, but sometimes still important.

Some atomic transitions of unusual interest have particularly small natural fractional widths. I am referring particularly to hyperfine transitions, which involve no change of the electronic states. The simplest for purposes of illustration, and the most important, is again atomic hydrogen. The energy levels in zero and very weak field are shown in Fig. 3-11.

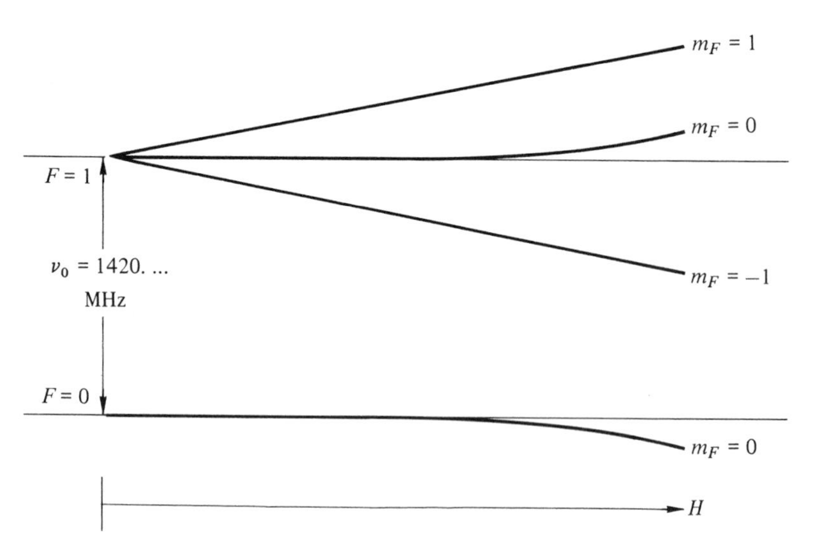

Fig. 3-11 Hyperfine structure of the ground state of the hydrogen atom, including Zeeman effect.

The transition frequency in very small fields between the $F = 1$, $m_F = 0$, and the $F = 0$, $m_F = 0$ states has been measured more accurately than any other number in physics by measuring the frequency of an atomic hydrogen maser operating between those levels [10]. The natural line width is on the order of cycles per second. Equation (3-43) gives a Doppler width of 20 kc/sec. Since the uncertainty in the measurement was about 2 parts in 10^{11}, the Doppler width could not have limited the accuracy of the experiment. What is the explanation? Motional narrowing of course!

The original proposal on how to defeat the Doppler shift was by Dicke [11], and it was phrased in the fm language. Imagine an atom in a one-dimensional box with perfectly reflecting walls (Fig. 3-12). Every $t = L/v$ seconds the particle reflects from the wall, and the frequency seen by an observer looking in the direction of motion changes abruptly from $\nu_0(1 + v/c)$ to $\nu_0(1 - v/c)$. Although the frequency modulation is square wave rather than sinusoidal, the spectrum will still be characterized by a modulation index m, which in this case is

$$m = \frac{2\pi\,\Delta\nu}{q} = \frac{\nu_0\,v/c}{v/2L} = 2\nu_0\,\frac{L}{c} = 2\,\frac{L}{\lambda}$$

Note that m is independent of v. The frequency spectrum will not differ qualitatively from the spectrum for sinusoidal modulation (Fig. 3-2). The sidebands are spaced by $v/L = q$. For the hydrogen maser experiment referred to, the hydrogen was confined to a bulb 15 cm in diameter. Although $m = \frac{3}{2}$ for this case—not even less than unity (since $\lambda = 21$ cm)—the sidebands are no closer than several thousand cycles away from the center frequency if the atoms in the beam came from an oven at room temperature, and, since the beam velocity is not monochromatic, the sidebands are broad, whereas the line at ω_0 is independent of v and has the natural line width.

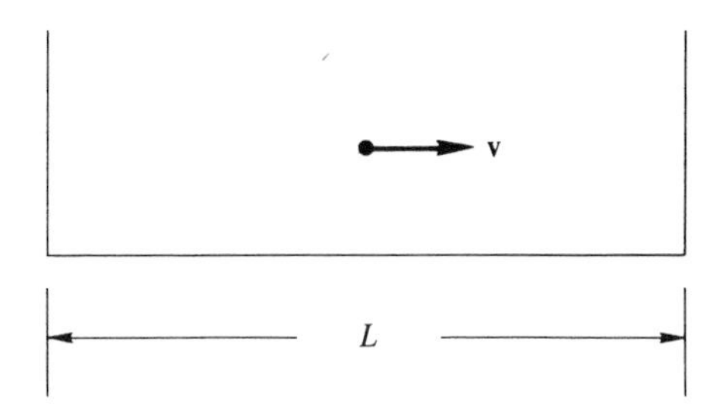

Fig. 3-12 One-dimensional box containing a single atom moving with velocity **v**. The walls are perfectly reflecting.

The method described works for hyperfine transitions because the wall collisions have a rather small effect on the hyperfine interaction. Also, a wall collision "lasts" for a short time compared with the period of the hfs transition. Dicke's original proposal was, in fact, a suggestion that a "buffer gas" be used in optical pumping experiments: that is, the sample container should be filled with rare gas atoms to a few millimeters pressure. The collisions between the atoms of interest, usually alkali atoms, and the rare gas atoms does not significantly disturb the hfs of the alkali atoms. But in making many collisions with the rare gas atoms, the resonant atom makes a random walk through the sample cell—a restrictive type of motion with many reversals and turnings—and is effectively confined into a small "box" of rare gas atoms roughly of the dimensions of the mean free path of alkali—rare gas collisions. The heavier rare gases, such as Kr and Xe, are used most effectively because "head-on" collisions between them and lighter atoms more nearly reverse the velocity of the light atoms. The collisions produce some frequency shift of the hyperfine transitions, with the magnitude and the sign of the shift dependent on the particular buffer gas used. Mixtures of buffer gases can consequently be found which produce no shift at all on the average.

Mössbauer effect

The Mössbauer effect is quite similar to the problem of "pressure narrowing" just discussed. The problem again is to observe the natural, lifetime-limited line breadth without the obscuring effects of Doppler broadening. The transition involved is between nuclear energy levels, and the γ-ray energies are from about 10 to over 100 keV, depending on the nucleus. Such a γ ray carries away substantial momentum, $h\nu/c$, and a free nucleus recoils with equal and opposite momentum, Mv. As a consequence, the γ-ray frequency seen by an observer is reduced from the magnitude it would have if the nucleus were infinitely massive by

$$\nu_0 - \nu = \nu_0 \frac{v}{c} = \nu_0 \frac{h\nu}{Mc^2} \tag{3-44}$$

The best known Mössbauer nucleus is ^{57}Fe, with a γ-ray energy of 14.4 keV. The fractional frequency shift given by Eq. (3-44) is thus approximately $(14.4 \times 10^3)/(10^9 \times 57) \simeq 0.25 \times 10^{-6}$. But the lifetime of the 14.4 keV state of ^{57}Fe is about 10^{-7} sec; the frequency equivalent to 14.4 keV is 3.5×10^{18} sec^{-1}. Therefore the natural fractional width is about 10^{-11}, and the frequency shift is several orders of magnitude larger than the line width. Consequently, a γ ray emitted by one free nucleus would not be absorbed by another one at rest, since the one at rest requires a photon Doppler shifted toward higher frequencies to conserve momentum in the absorption.

One way of describing the Mössbauer effect is simply to say that both source and absorber nucleus are placed in a solid, and the macroscopic mass of the crystal is the M of Eq. (3-44), so the frequency shift is negligible. Let us regard that explanation as insufficient, or at least incomplete, and take a closer look at a solid. The Einstein model may not be terribly good quantitatively, but it contains enough useful truths to serve us here.

In the Einstein model, each atom or molecule vibrates independently in a simple harmonic oscillator potential with fundamental frequency ν_E. If the nucleus is regarded as a classical radiator, the spectrum of the emitted radiation will simply be one of the spectra of Fig. 3-3, and q in that figure would be $2\pi\nu_E$. If the modulation index m is less than, or on the order of, unity, then the central line at ν_0 contains most of the radiated power. It is interesting to compute the degree of harmonic oscillator excitation for $m = 1$ in this case, since that value of m represents some sort of dividing line between power mainly in the central line and power mainly in the sidebands. To find the depth of modulation, we need the maximum velocity of the oscillating atom. Assume we can use the correspondence principle limit (which will be good for high n)

$$nh\nu_E = \tfrac{1}{2}Mv_{\max}^2$$

From Eq. (3-44), the frequency deviation is

$$\Delta\nu = \nu_0\left(\frac{2nh\nu_E}{Mc^2}\right)^{1/2}$$

and

$$m = \nu_0\left(\frac{2nh}{Mc^2\nu_E}\right)^{1/2} \tag{3-45}$$

Set $m = 1$, solve for n:

$$n = \tfrac{1}{2}\left(\frac{Mc^2}{h\nu_0}\right)\left(\frac{\nu_E}{\nu_0}\right) \tag{3-46}$$

The ratio in the first bracket we recognize from Eq. (3-44) to be the inverse fractional shift of the Doppler-shifted line; it is about 4×10^6 for ^{57}Fe. A reasonable value for ν_E is 10^{13} Hz, so ν_E/ν_0 is about 0.3×10^{-5}, and hence $n = 10$. This result means that in an Einstein solid most nuclear γ-ray emissions from ^{57}Fe will be unshifted at temperature T, where $3kT < nh\nu_E$. With the numbers we have been using, for $m = 1$, $T \sim 200°K$; that is, the Mössbauer effect is a low temperature phenomenon. That is a qualitatively correct conclusion, although, as our carefully chosen numbers also suggest, the phenomenon can easily be observed at room temperature also, at least for ^{57}Fe.

The same qualitative conclusions are reached for the Debye model of lattice vibrations. The sidebands are no longer displaced by discrete frequencies ν_E on either side of ν_0, but are spread continuously, and hence unobservably, over about the same frequency region as the sidebands appear in the Einstein model. The central, sharp line in the broad background is sometimes known as the zero-phonon line, since it corresponds to γ-ray emission with no change in the state of lattice vibrations. In the Einstein model, the various sidebands at $\pm\nu_E$, $\pm 2\nu_E$, and so forth, would, in this language, correspond to simultaneous emission $(-)$ or absorption $(+)$ of one, two, and so forth, phonons.

The correctness of the point of view used here with the Einstein model was demonstrated experimentally by Ruby and Bolef [12] (see Frauenfelder [13] for a reprint of that paper). They placed a source on a quartz transducer vibrating at 20 MHz and actually observed the various sidebands grow as the modulation index was increased. (See also a recent letter to *Physical Review Letters* by Heiman, Pfeiffer, and Walker [14].)

If the student had expected to see just how in the zero-phonon line the recoil momentum was taken up by the crystal as a whole, he is no doubt feeling disappointed by this time. Indeed, the point is quite subtle, and we will continue to disappoint by not pursuing the various points of view which help make the answer at least more palatable. The reprint volume by Frauenfelder [13] is again recommended for further work. Note that the first paper reprinted there is the one by Dicke [11] referred to in the previous section.

In the more complete theories, the Mössbauer or zero-phonon line is attenuated at nonzero temperatures by an exponential function called the Debye–Waller factor. It was known under that name by a generation of X-ray crystallographers before the Mössbauer effect was discovered. The elementary theory of X-ray diffraction—the one which is essentially physical optics, and which yields the Bragg law—usually does not mention the problem of conservation of momentum in the coherent scattering of X rays by a crystal lattice. Figure 3-13 reminds you of the geometry of Bragg reflection and defines the Bragg angle θ. Each photon in the diffracted beam has changed its momentum by $2(h\nu_0/c)\sin\theta$ normal to the atomic planes from which the diffraction occurs. Yet the diffracted beam is still at the same wavelength or frequency as the incident wave to very

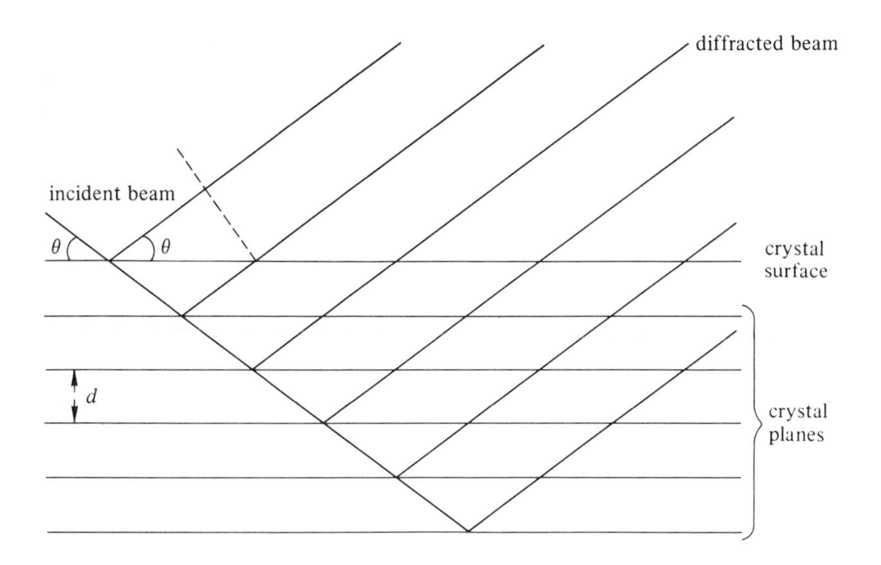

Fig. 3-13 Geometry of X-ray diffraction. Diffracted beam obeys the Bragg law $n\lambda = 2d \sin \theta$.

high precision.[4] The explanation is the same as for the Mössbauer effect. The absence of a Doppler shift in the diffracted beam means the recoiling scatterer must be very massive, so that it recoils with very low velocity and produces negligible Doppler shift of the scattered beam.

The invocation of frequency modulation to explain the effects of atomic vibrations is also the same as in the Mössbauer effect. One must only realize that the frequency of radiation scattered from a moving scatterer is Doppler shifted by the amount given by Eq. (3-44). The rest of the explanation follows exactly the same reasoning as for the Mössbauer effect, with the following difference. Even for the Einstein model, the sidebands are not likely to be prominent, since X-ray diffraction is a *coherent* scattering from the large number of scatterers in the crystal. In the Einstein model, the sidebands are produced by the independently vibrating atoms, so the phases of scattered radiation in the sidebands, as well as the direction of scattering, will wash out the sideband structure. Only the central frequency line at ν_0 is scattered coherently, in phase as well as spatially; therefore, only it contributes to the Bragg peak.

[4] The process of detection of the diffracted beam, a photographic plate or a Geiger–Müller tube, is not sensitive to a change of wavelength, as is the process of reabsorption of the γ ray by a Mössbauer nucleus in its ground state. The constancy of ν_0 in the X-ray diffraction experiment may be shown by using the diffracted beam as the incident beam in a subsequent diffraction experiment on a crystal of the same type. The Bragg angles for the two experiments are the same within very small experimental error.

3-6. SUMMARY AND LITERATURE SURVEY

We have presented in this chapter the bare bones of a semiquantitative theory (if the word is not too strong) of line narrowing by nuclear motion, or perhaps better, line narrowing by frequency modulation. Since our treatment did not develop even the beginnings of the apparatus necessary for the elegant and powerful general theory, the purely theoretical development was necessarily brief. It is hoped that the range of examples compensates for the brevity of the initial treatment. As we warned at the beginning, some examples might strike readers knowledgeable in other fields as somewhat strained in their application of this chapter's one idea. Thus, to regard the Mössbauer effect as an example of the same thing that causes the proton resonance to be narrow in water is regarded as a curious parochialism by the sophisticates of Mössbauery; and so it may be, but the application illuminates both fields.

The ideas in this chapter were well formed and well understood very quickly after the first magnetic resonance observations. The 1948 thesis by Bloembergen [3] and the famous BPP paper [15] based on it contain all the ideas and many of the applications we have presented here. Scalar spin coupling and the effects of rapid relaxation on it were discussed in Gutowsky *et al.* [6]. The paper merits study for the beautiful proof that scalar coupling among like spins has no effect on the macroscopic transverse magnetization, and for the detailed calculation of the "exchange narrowing" model. The random walk language, which lead to our Eq. (3-4), was published by Slichter [16] after several years service at the chalkboard in the laboratory. Some of the more difficult examples of its application in that paper can profitably be skipped, however.

We have not mentioned a great many of the very well-known examples of motional narrowing. One of the most celebrated is the study of hindered rotation in solids, such as the reorientation of the ammonium molecule in NH_4Cl and related substances. Some of the effects involving hindered rotation, as it is sometimes called, are still not thoroughly understood, and are the subject of some current work. The early work by the Harvard group is discussed by Pake [17].

Many of the interesting conundrums and "tricky" experiments associated with the Mössbauer effect are explored in the introduction and reprint volume by Frauenfelder [13]. A very clear, delightfully written discussion of frequency modulation, to which the present treatment owes much, may be found in Arguimbau [18].

We leave the reader with one last teaser. Is the Ramsey split-field molecular beam resonance experiment really an example of artificial motional narrowing? The answer will be found somewhere in Chapter 5.

Problems

3-1. Suppose one is doing a magnetic resonance experiment in a system in which the local field is entirely transverse to the applied field. Assume that the x and y transverse components fluctuate independently with the same correlation time τ_c while taking on the two values $+h_L$ and $-h_L$. Show both by Eqs. (3-28) and (3-29), and by a random walk argument, that $T_1 = T_2/2$. (This model of a fluctuating local field is an artificial one concocted to give the result $T_1 < T_2$, which is sometimes seen in physical systems, such as in nuclear magnetic resonance in ferromagnets.)

3-2. The line shape of the nuclear magnetic resonance signal of water in a particular magnet is a double-peaked curve, with the peak separation measured to be 10^{-2} G. Assume the sample can be rotated about some axis such that the nuclei contributing to one of the peaks will move to a position in the magnetic field so that they contribute to the resonance at the frequency of the other peak. Find the sample spinning frequency such that the double-peaked line shape just becomes a single peak. Find the spinning frequency such that the spinning sidebands are 4×10^{-2} G from the center line. How intense are they relative to the main line?

3-3. Consider a hypothetical molecule BE_2, $\gamma_B \neq \gamma_E$, and $I_B = \frac{3}{2}$, $I_E = \frac{1}{2}$. The E atoms in the molecule are chemically equivalent, such as the hydrogen atoms are in H_2O. There is a coupling of the form $A\mathbf{I}_B \cdot (\mathbf{I}_E)$ between the B atom and each E atom. Sketch the high field magnetic resonance spectrum observed for each nucleus as a function of frequency (constant field). Include such details as the relative intensities and frequency splittings. Find an estimate for T_{1B} such that the E spectrum collapses to a single line.

3-4. Consider a single line ^{57}Fe Mössbauer source placed on a transducer, as in the experiment of Bolef and Ruby [12]. Assume that the source undergoes simple harmonic motion at 20 MHz. Find the amplitude of the vibrations if the Mössbauer spectrum is characteristic of a modulation index $m = 2$.

3-5. If a buffer gas instead of a box is used to reduce line width broadening due to the Doppler effect, show that a reasonable estimate of the modulation index would be Λ/λ, where Λ is the mean free path between collisions and λ the wavelength of the hyperfine transition. Show that the line width of the transition can be written $\Delta\nu \sim D/\lambda^2$, where D is the diffusion constant of the atom of interest in the buffer gas ($D = v\Lambda/3$, where v is the average speed of the atom).

References

1. C. P. Slichter, *Principles of Magnetic Resonance*, Harper & Row, New York (1963).
2. R. P. Feynman, R. B. Leighton, and M. Sands, *The Feynman Lectures on Physics*, Addison-Wesley Publishing Co., Reading, Massachusetts (1965), vols. 1, 2, 3.

3. N. Bloembergen, *Nuclear Magnetic Relaxation*, W. A. Benjamin, Inc., New York (1961).
4. *Handbook of Mathematical Functions*, M. Abramowitz, and I. A. Stegun, Eds., National Bureau of Standards, Washington, D.C. (1964). See formulas 9.1.42 and 9.1.43, p. 361.
5. F. Reif, *Fundamentals of Statistical and Thermal Physics*, McGraw-Hill Book Company, New York (1965), pp. 583ff.
6. H. S. Gutowsky, D. W. McCall, and C. P. Slitchter, *J. Chem. Phys.* **21**, 279 (1953).
7. J. S. Waugh, L. M. Huber, and U. Haeberlin, *Phys. Rev. Letters* **20**, 180 (1968).
8. J. H. VanVleck, *Phys. Rev.* **74**, 1168 (1948).
9. P. W. Anderson, *J. Phys. Soc. Japan* **9**, 316 (1954).
10. S. B. Crampton, D. Kleppner, and N. F. Ramsey, *Phys. Rev. Letters* **11**, 338 (1963).
11. R. H. Dicke, *Phys. Rev.* **89**, 472 (1953).
12. S. L. Ruby and D. I. Bolef, *Phys. Rev. Letters* **5**, 5 (1960).
13. H. Frauenfelder, *The Mössbauer Effect*, W. A. Benjamin, Inc., New York (1962).
14. N. D. Herman, L. Pfeiffer, and J. C. Walker, *Phys. Rev. Letters* **21**, 93 (1968).
15. N. Bloembergen, E. M. Purcell, and R. V. Pound, *Phys. Rev.* **73**, 679 (1948).
16. D. Pines and C. P. Slichter, *Phys. Rev.* **100**, 1014 (1955).
17. G. E. Pake, *Solid State Physics*, F. Seitz and D. Turnbull, Eds., Academic Press Inc., New York, vol. 2 (1956).
18. L. B. Arguimbau, *Vacuum Tube Circuits and Transistors*, John Wiley & Sons, Inc., New York (1956).

CHAPTER 4

Nuclear Magnetic Resonance in Solids

Just what a "solid" is in this chapter must be carefully defined, since many solids display liquidlike behavior, and are more correctly described by Chapter 3, as far as nuclear magnetic resonance is concerned. For our purposes, a solid is a collection of atoms spaced closely enough so the density is on the order of grams per cubic centimeter, and for which $\omega\tau_c \gg 1$. The inequality expresses the requirement that the magnetic moments stay fixed long enough for the full effect of the spin-spin interaction (which may or may not be dipole-dipole) to be manifested. The definition of a solid is made here in this way to emphasize the point that nuclear motion can be so rapid for solids near but below the melting point that the substance may behave as a liquid in the sense that it may satisfy the Bloch equations. To emphasize the point, we shall discuss the type of solid that is often referred to as a *rigid-lattice* solid. Spins imbedded in the rigid lattice do not obey the Bloch equations. The line widths are determined by a lattice sum of the spin-spin interaction, and the shapes are not usually Lorentzian (and, as a corollary, the free induction decay is not exponential). The T_1 processes are usually quite independent of the mechanisms giving rise to the line width, which are still roughly described by a parameter, T_2, the transverse relaxation time. Furthermore, if one roughly characterizes the line width $\delta\omega$ by its inverse, $T_2 = 1/\delta\omega$, then in solids it is most frequently true that $T_1 \gg T_2$. All these facts add to make for tremendous variety and complication in magnetic resonance experiments in solids.

In this chapter we shall introduce some of this variety in the following way. Sections 4-1 and 4-2 deal with rigid-lattice line widths and the characterization of the shapes of resonance lines. Section 4-3 explores some of the thermodynamics of spin systems. Section 4-4 discusses the nuclear quadrupole interaction, one feature of the chapter not necessarily dealing specifically with a solid. The last section discusses qualitatively or semiclassically some of the common physical mechanisms responsible for spin-lattice relaxation.

4-1. RIGID-LATTICE HAMILTONIAN

Consider a regular lattice with, say, nuclei of spin I_1, I_2, ..., and gyromagnetic ratios γ_1, γ_2, ..., on lattice sites. In the presence of the static field, the quantum mechanical statement of the problem is contained in the Hamiltonian

$$\mathscr{H} = \mathscr{H}_Z + \mathscr{H}_{II} \tag{4-1}$$

where $\mathscr{H}_Z$, the Zeeman term, describes the interaction of the spins with the external field in the z direction:

$$\mathscr{H}_Z = -\sum_{i\nu} \gamma_\nu \hbar H_0 I_{i\nu}^z \tag{4-2}$$

The sum over ν is over the types of nuclei in the sample; the index i is summed over the lattice. An example is the alkali halide NaF. There are two types of nuclei, ^{23}Na, with spin $\frac{3}{2}$, and ^{19}F, with spin $\frac{1}{2}$. If we designate by N the number density of molecules of NaF, Eq. (4-2) may be written

$$\mathscr{H}_Z = -(\gamma_{\text{Na}} \hbar I_{\text{Na}}^z + \gamma_{\text{F}} \hbar I_{\text{F}}^z) N H_0 \tag{4-3}$$

The sum over i gives the factor N; the arrangement of spins on the lattice does not matter because each spin interacts independently with the external field, and the interaction does not depend on the spin's position. The two parts of Eq. (4-3) are completely independent, and specify so far two independent problems.

Before moving on to consider the second term in Eq. (4-1), the interaction part, we are wise to get a complete grasp of the structure of the energy levels of Eq. (4-3). Since that equation does have two independent parts, we may examine either one of them. Figure 4-1 illustrates the Zeeman energy levels of a spin system of spin I in the field H_0. The quantity M ranges between $-NI$ and $+NI$, and the range of energies is from $-\hbar\omega_0 NI$ to $+\hbar\omega_0 NI$. Since there are $2NI + 1$ such energy levels, it is clear that there must be great degeneracy, for the number of states is $(2I + 1)^N$. For example, the degeneracy of state M is

$$\frac{2^N N!}{(N - 2M)!(N + 2M)!} \quad \text{for} \quad I = \tfrac{1}{2}$$

It must seem arbitrary to introduce gratuitously this immense complication into a fundamentally simple problem, since all the operators $I_{i\nu}^z$ are independent and the Zeeman Hamiltonian contains no terms in the product

$\hbar\omega_0(NI)$
$\hbar\omega_0(NI-1)$
$\hbar\omega_0(NI-2)$
$\hbar\omega_0(NI-3)$
$\hbar\omega_0/2$
$-\hbar\omega_0/2$
$-\hbar\omega_0(NI-2)$
$-\hbar\omega_0(NI-1)$
$-\hbar\omega_0(NI)$

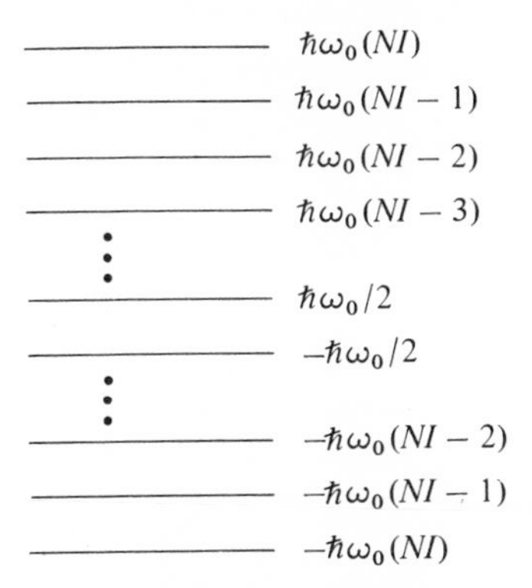

Fig. 4-1 Energy levels of N spins of spin I in a magnetic field $H_0 = \omega_0/\gamma$. (Figure drawn for N, an odd number, and I, an odd half-integer.)

of two different operators $I_{i\nu}^z$ and $I_{j\nu'}^z$, say. But the $\mathscr{H}_{II}$ part of Eq. (4-1) does contain such products, and we must prepare for the consequences. Figure 4-2 illustrates the spin states for the second most trivial possible example, that of two identical spins of $I = 1$. Note there are $(2I + 1)^2 = 3^2 = 9$ states; all but the extreme energy levels are degenerate.

If the magnetic field H_0 is large enough, the term $\mathscr{H}_Z$ in Eq. (4-1) may be made much the larger of the two, and the energy-level diagram of Fig. 4-2 will be nearly correct even in the presence of $\mathscr{H}_{II}$. This interaction term is bilinear in the different spins, and, in general, quite complicated. For example, if it takes the form of the dipolar interaction between identical spins, it is

$$\mathscr{H}_{\text{dip}} = \sum_{i>j} \frac{\gamma_i \gamma_j \hbar^2}{r_{ij}^3} \left(\mathbf{I}_i \cdot \mathbf{I}_j - \frac{3\mathbf{I}_i \cdot \mathbf{r}_{ij} \mathbf{I}_j \cdot \mathbf{r}_{ij}}{r_{ij}^2} \right) \tag{4-4}$$

The sum is a double one over the entire spin systems I_i, I_j, where $r_{ij} = |r_i - r_j| = |r_{ij}|$ is the distance between spins i and j. The notation of Eq. (4-4) is actually quite compact. To see the real consequences of Eq. (4-4), it is necessary to write out all the scalar products in Cartesian

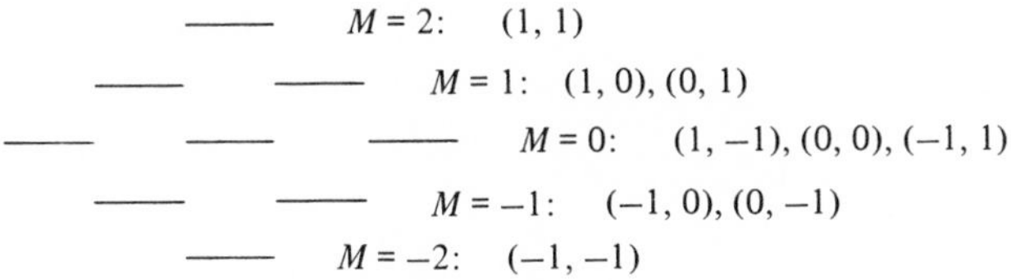

$M = 2$: (1, 1)
$M = 1$: (1, 0), (0, 1)
$M = 0$: (1, −1), (0, 0), (−1, 1)
$M = -1$: (−1, 0), (0, −1)
$M = -2$: (−1, −1)

Fig. 4-2 Energy levels and values of M for two identical spins of $I = 1$ in a magnetic field.

coordinates, rewrite in terms of the so-called raising and lowering operators, $I_j^+ = I_j^x + iI_j^y$, $I_j^- = I_j^x - iI_j^y$, and regroup. (Here, $i = \sqrt{-1}$.) The regrouping is determined by the observation that I_j^+ and I_j^- are operators having matrix elements between states that differ by $m_I = \pm 1$ [see Eq. (3-39), and the discussion surrounding it]. In the result, all possible combinations of pairs of spin operators occur between each pair of spins i and j. We can classify the pairs by the difference ΔM between the states of the Zeeman Hamiltonian, Eq. (4-23), between which the pairs of spin operators have matrix elements. The types of operator pairs for $\Delta M = 0$, ± 1, ± 2 are shown in Table 4-1.

Table 4-1
Classification of Spin Operators
in the Dipole-Dipole Hamiltonian

ΔM	Spin operators
0	$I_i^z I_j^z, I_i^+ I_j^-, I_i^- I_j^+$
± 1	$I_i^z I_j^+, I_i^z I_j^-, I_i^+ I_j^z, I_i^- I_j^z$
± 2	$I_i^+ I_j^+, I_i^- I_j^-$

If we work in the Zeeman representation, that is, the representation of Fig. 4-1 in which $M = \sum_i m_i$ is a good quantum number, and if we regard the spin-spin interaction as a perturbation, then we see from Table 4-1 that part of the dipole interaction is diagonal in that representation and part of it is not. It is important to call attention to the fact that the scalar exchange interaction between two spins, $\mathbf{I}_i \cdot \mathbf{I}_j$, is entirely diagonal in the Zeeman representation, since it is made up of a linear combination of the three pairs in the $\Delta M = 0$ line of Table 4-1. The operators in this line produce what is called a *secular perturbation* on the energy-level diagram; the other lines contain *nonsecular* operators. The distinction is very important—so much so that, under the conditions $\mathscr{H}_Z \gg \mathscr{H}_{II}$, one may ignore the remaining lines of Table 4-1. So that the point is perfectly clear, consider the $M = 1$ states in Fig. 4-2. The operator $I_1^z I_2^z$ will shift each state of that degenerate pair upward by the coefficient of $I_1^z I_2^z$, an amount roughly given by $(\gamma\hbar)^2/r^3$, in the case that $\mathscr{H}_{II}$ is the dipolar Hamiltonian. The effect of the other two terms is to *lift* the degeneracy, splitting the states by an amount proportional to the coefficient of the operators, which again is roughly $(\gamma\hbar)^2/r^3$ if $\mathscr{H}_{II}$ is dipolar. This result can be seen by students who have not been introduced to the formalism of degenerate perturbation theory by pointing out the analogy to singlet and triplet states of two-spin systems in atomic physics. The lifting of the degeneracy by $I_1^+ I_2^- + I_1^- I_2^+$ can be seen directly by going to the answer: the wavefunctions that are diagonal in $I_1^+I_2^- + I_1^-I_2^+$ are not the pair $|m_1m_2\rangle = |01\rangle$ and $|10\rangle$, but the linear combinations $|\chi_1\rangle = [|01\rangle + |10\rangle]/\sqrt{2}$, $|\chi_2\rangle = [|01\rangle - |10\rangle]/\sqrt{2}$. Compute the matrix elements $\langle\chi_1| I_1^+I_2^- + I_1^-I_2^+ |\chi_1\rangle$ and $\langle\chi_2| I_1^+I_2^- + I_1^-I_2^+ |\chi_2\rangle$; they are different.

On the other hand, the other operator products of Table 4-1 have matrix elements between states that differ in energy by $\pm\gamma\hbar H_0$ for the operators in the $\Delta M = \pm 1$ line, and by $\pm 2\gamma\hbar H_0$ for the operators in the $\Delta M = \pm 2$ line. Their effect on the energy levels is approximately $\gamma\hbar {h_L}^2/H_0$, where $h_L \sim \gamma\hbar/r^3$ is a measure of the local dipolar field. But we ignore them for another reason beside their smallness. We are interested in computing the effect of the dipole interaction on the line shape of the resonance. The resonance experiment is done by applying a transverse rf field $H_1 \cos \omega t$ and measuring, if χ'' is being detected, the power lost to the sample at $\omega \simeq \omega_0$. The rf field represents an additional perturbation to Eq. (4-1), smaller than $\mathscr{H}_{II}$ if there is no saturation, given by

$$\mathscr{H}_{\text{rf}} = -\hbar H_1 \cos \omega t \sum_{i\nu} \gamma_\nu I_{i\nu}^x \tag{4-5}$$

by direct analogy to Eq. (4-2). The important thing to realize is that one looks at power absorption near and at the Larmor frequency. The operators on the $\Delta M = \pm 1, \pm 2$ lines of Table 4-1, in addition to producing a second-order change in the energy (i.e., a change proportional to ${h_L}^2$), also produce a first-order change in the wavefunction. We use Fig. 4-2 as an example again. The state (1, 1) corresponding to $M = 2$ is not "pure" but has mixed in it small amounts of each of the states of $M = 1$ and $M = 0$. The relative size of the admixture is on the order of h_L/H_0, but the admixture, however small, allows matrix elements to exist between the $M = 2$ and $M = 0$ states. These states are separated by $2\hbar\omega_0$ in energy, and absorption would be observed at $2\omega_0$, down in intensity by $(h_L/H_0)^2$ from the main line at ω_0. We somewhat arbitrarily restrict ourselves to looking for absorption near ω_0, so we do not wish to include the little bump at $2\omega_0$ as part of our line shape calculation. We can see the "zero frequency" absorption in the two-spin example of Fig. 4-2 by examining the admixtures into the two $M = 1$ states of the $M = 0$ and $M = 2$ wavefunctions, and by noting that there is thus a nonzero matrix element of I^x between the pair of perturbed wavefunctions of $M = 1$.

The preceding argument has justified the process known as "truncation" of the dipole-dipole interaction. We throw away the nonsecular terms of Eq. (4-4) and keep only those of the $\Delta M = 0$ line of Table 4-1. For completeness, we write out the remainder of the dipole interaction in spherical coordinates Eq. (4-6). The angles and distances are defined in Fig. 4-3.

$$\mathscr{H}_{ij} = \frac{\gamma^2\hbar^2}{r_{ij}^3}\,[I_i^z I_j^z - \tfrac{1}{4}(I_i^+ I_j^- + I_i^- I_j^+)](1 - \cos^2\theta_{ij}) \tag{4-6}$$

The argument has so far been rather lengthy and qualitative. Let us summarize what has been done and generalize a little bit. We want to compute the absorption line shape of a resonance experiment in the presence of an external field H_0 that is much larger than the spin-spin interaction. Assume, to begin with, that only one spin species is present. The states of the system may be described in terms of the states of $\mathscr{H}_z$, which are labeled according to the quantum number $M = \sum m_i$, where M ranges between $-NI$ and $+NI$. Except for the two extreme levels, all these levels are highly degenerate. The degeneracy is removed, and the ladder of discrete levels is changed to a quasi-continuum of levels by the secular part of the spin-spin interaction, which is the part that has matrix elements only between states of the same M. The most important effect of the rest of the spin-spin interaction is to allow weak transitions between states differing in frequency by $\omega = 0$ and $2\omega_0$. Since the absorption line shape will be examined only near $\omega = \omega_0$, the spin-spin Hamiltonian is *truncated* by excluding all nonsecular terms.

Now we shall be able to see how to treat the spin-spin interaction between unlike spins. Suppose we are observing the resonance of spins I that interact with spins S. The complete Hamiltonian has the two Zeeman terms and the three spin-spin terms:

$$\mathscr{H} = \mathscr{H}_{IZ} + \mathscr{H}_{SZ} + \mathscr{H}_{II} + \mathscr{H}_{SI} + \mathscr{H}_{SS} \tag{4-7}$$

Everything we have said previously about a single spin is still true as far as $\mathscr{H}_{IZ} + \mathscr{H}_{II}$ is concerned. The terms $\mathscr{H}_{SS}$ and $\mathscr{H}_{SZ}$ will have no direct effect on our resonance of the I spins, and an indirect effect only under very

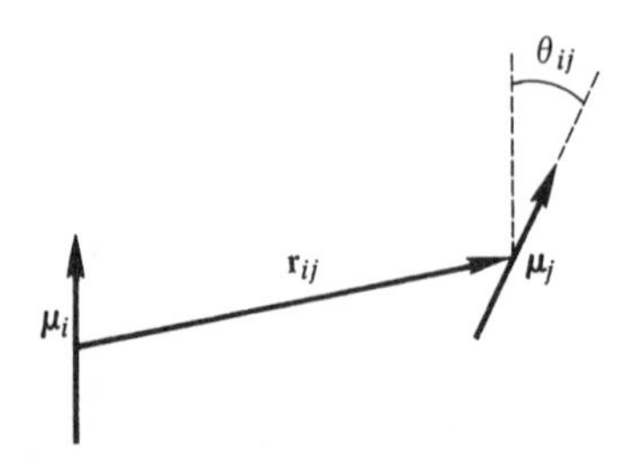

Fig. 4-3 Geometry for interacting dipoles.

special circumstances. The coupling term $\mathscr{H}_{SI}$ has spin operators I_i^x, I_i^y, I_i^z in it, and contributes to the line width and shape. But we must truncate the complete $\mathscr{H}_{SI}$ term to include only the secular part, in accordance with the procedure described. That leaves only the $I_i^z S_k^z$ type terms, and does not include terms such as $I_i^+ S_k^-$ and $I_i^- S_k^+$, because these operators connect states of different energy. The energies differ by $|(\gamma_I - \gamma_S)| \hbar H_0$, and as long as $|\gamma_I - \gamma_S| H_0$ is larger than the strength of the spin-spin interaction—that is, as long as the resonance lines for the I spins and S spins do not overlap—then $I_i^z S_k^z$ terms may be kept in the truncation. We also see that if the γ's are roughly comparable, interactions between like spins are more effective in broadening resonance lines than interactions between unlike spins because of the energy conserving $I_i^+ I_j^-$ mutual spin flip terms in the former case.

4-2. THE METHOD OF MOMENTS

It must be getting apparent by this time that although we speak of our goal as the calculation of the shape of the resonance line, we have not admitted that it can be done. It is, in fact, true that the problem in general has not been solved, but that statement is true to some degree about most interesting problems in physics. In the case of the resonance line shape, some reasonably accurate solutions have, in fact, been obtained within the last decade. It is instructive, but disappointing, to realize that there does not seem to be too much gained by solving the line shape problem under most circumstances. There are two major reasons for this fact. One is that the methods of solution so far used are difficult to apply, in the sense that the calculations are long, tedious, susceptible to error, and not particularly revealing of the physics of the problem. A more important reason is that in the first decade of activity in the field of magnetic resonance, powerful, general, and yet quite usable methods for discussing line shapes were developed and were universally used and understood. They served the needs of the physics so well that they have not been displaced.

The principal technique for describing and calculating the shape of a resonance line in a rigid-lattice solid is the method of moments, developed in 1948 by Van Vleck [1]. Unfortunately for the expository author on the level of this book, a detailed exposition of the method requires rather more familiarity with the *apparatus* of quantum mechanical calculation than could reasonably be expected. We shall therefore have to content ourselves with a few definitions, a statement of results, and an attempt to convey the usefulness of the method by discussing a few of the standard examples. Rather more effort will be made to touch bases for their own sake than has been done previously.

Let us describe a resonance line by the normalized function $g(\nu)$ first introduced following Eq. (2-46). We define the nth moment of $g(\nu)$ by the equation

$$\langle \nu^n \rangle = \int_0^\infty \nu^n g(\nu)\, d\nu \tag{4-8}$$

If $g(\nu)$ is symmetric about the resonance frequency, then $\langle \nu \rangle = \nu_0$. The moment relative to the center of the resonance line is defined by

$$\langle \Delta\nu^n \rangle = \int_0^\infty (\nu - \nu_0)^n g(\nu)\, d\nu \tag{4-9}$$

For $n = 2$, the binomial in the integrand may be expanded, and we obtain the expression

$$\langle \Delta\nu^2 \rangle = \langle \nu^2 \rangle - 2\nu_0 \langle \nu \rangle + \nu_0^{\,2}$$

or

$$\langle \Delta\nu^2 \rangle = \langle \nu^2 \rangle - \nu_0^{\,2} \tag{4-10}$$

since $\langle \nu \rangle = \nu_0$.

The moments of a curve do provide a way, admittedly indirect, to describe it. For example, the curve given by

$$g(\nu) = \frac{1}{\sigma\sqrt{2\pi}} \exp\left[\frac{-(\nu - \nu_0)^2}{2\sigma^2}\right] \tag{4-11}$$

is called a Gaussian curve. Its width is determined by the parameter σ. One may check in a table of integrals that not only $\langle \Delta\nu^2 \rangle = \sigma^2$, but also $\langle \Delta\nu^{2n} \rangle = [1 \cdot 3 \cdot 5 \cdot \cdots \cdot (2n - 1)]\sigma^{2n}$. Another single parameter resonance curve we have encountered was Lorentz curve, Eq. (2-44). It is easy to see that since it dies off as $(\nu - \nu_0)^{-2}$ and the multiplier in the integrand of Eq. (4-9) increases as $(\nu - \nu_0)^{2n}$, all moments for $n \geq 2$ are infinite. That is not too alarming because there are good physical reasons for cutting off a long-tailed curve such as Eq. (2-44) below unphysical frequencies corresponding to energies beyond the realm of the physical system.

The advantage of the moment method is that Van Vleck has provided an exact formula for its calculation. Although the details are beyond the scope of this book, we shall describe in words the essence of the calculation. The most complete introduction to the details is to be found in Slichter [2].

The first step is to develop a formal expression for χ'' from quantum statistical mechanics for the system of Zeeman plus spin-spin interactions. The second step, taken by Van Vleck, is to write that formal expression as the trace of the square of the commutator of the Hamiltonian and the operator I^x. In that mouthful of jargon the key word is "trace," which is the term for the diagonal sum of a matrix. Since that diagonal sum is independent of the representation that is used, Van Vleck's expression may be evaluated in the Zeeman representation of Fig. 4-1, even though M is not a good quantum number in the presence of spin-spin interactions. The usefulness of the moments technique rests on that observation. The Hamiltonian used in the moment expression is the Zeeman plus the *truncated* spin-spin Hamiltonian, for reasons we have already discussed at length.

We conclude these few words on moments by mentioning some standard results that are so well known they form part of the language of magnetic resonance.

1. The second moment calculated by using the entire dipole Hamiltonian is 10/3 as great as it is by using the truncated one. The reason is that the $\Delta M = 0$, and $\Delta M = 2$ (and higher) contributions, although weak, are weighted by ${\nu_0}^2$ in the second moment calculation. Hence, they contributed roughly the same as the center line at ν_0. This is an important observation if the resonance is done in small or zero field.
2. A scalar exchange interaction between like spins does not contribute to the second moment, but it does increase the fourth moment. Consider a sample with a single spin species the line shape and width of which is determined entirely by the dipole interaction. Such lines are typically Gaussian or somewhat more "square shouldered" than Gaussian. If one were to "turn on" the exchange interaction, the line shape would change by developing a narrower center and broadened wings. Superficially it would become more Lorentz-like. The wings broaden to increase the fourth moment, the center peaks to maintain the second moment constant. The fourth moment is, of course, more sensitive to the wings than the second moment. Figure 4-4 illustrates the statement. Such a line is often called "exchange narrowed," although use of the term "narrow" may seem arbitrary. Reference to our discussion of the phenomenon in Chapter 3 illuminates the reason for the line shape change. The broad, weak wings are the sidebands in our fm model for the case of small modulation index. The increased intensity of the center agrees with the picture developed by that model.
3. The second moment is independent of nuclear motion. The discussion of this observation is very much the same as for the case of exchange narrowing. It seems strange at first that the very narrow proton resonance in water ($\sim\frac{1}{3}$ Hz) has the same second moment as does the

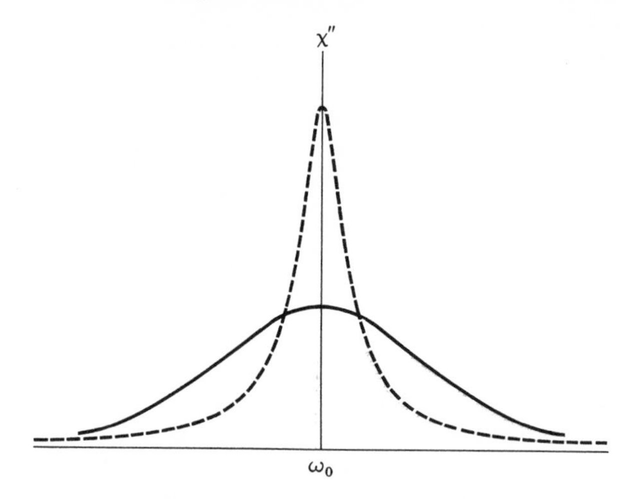

Fig. 4-4 Solid curve: Typical dipolar line shape in a solid. Dashed curve: Exchanged-narrowed line shape in same lattice [after G. Pake, *Paramagnetic Resonance*, W. A. Benjamin, Inc., New York (1962)].

proton resonance in ice ($\sim$50 kHz). Of course, it really does not; the observed line of $\frac{1}{3}$ Hz in width is not the entire line because the sidebands are so weak and spread out as to be unobservable. This example demonstrates the fact that it is wise to be very careful in comparing calculated second moments with apparent experimental ones. Broad, weak wings of resonance lines merge very quickly into noise and are quite invisible.

The calculation of a second moment ultimately reduces to the evaluation of a lattice sum—that is, the sum over the positions on the lattice of functions of angles and distance. The second moment is structure sensitive; it is also a function of the orientation of external field relative to the crystal directions. That is, the line width and shape are not isotropic in general but vary as the crystalline orientation is changed in the field.

4-3. THERMODYNAMICS OF SPIN SYSTEMS

Consider the expression for the energy of N spins in a large magnetic field, Eq. (2-20).

$$E = -\frac{N\gamma^2\hbar^2 I(I+1)H_0{}^2}{3kT} \tag{2-20}$$

This expression represents the main energy of the spin system as long as $\gamma^2 H_0{}^2 \gg (2\pi)^2\langle\Delta\nu^2\rangle$. (The second moment is a very convenient quantitative measure of the mean square internal field.) We would now like to view E as the internal energy of a thermodynamic system of spins in internal equilibrium at temperature T. In Eq. (2-20), of course, T was the lattice temperature. We acquire greater understanding realizing that as long as $T_2 \ll T_1$, the system of N spins interacts within itself more strongly than with the lattice and will come to internal equilibrium more rapidly than it will come to equilibrium with the lattice.

The basic understanding of the mechanism by which a spin system comes to internal equilibrium is due to Bloembergen [3]. The mechanism is called spin diffusion. Any diffusion process involves the transport of something, driven by a concentration gradient. In this case, the "something" being transported is spin magnetization, and the gradient is a spatial gradient in magnetization. The magnetization does not diffuse by physical movement of the nuclei or electrons which have the magnetic moments. Rather it proceeds by mutual spin flips of like moments via the term in the dipole interaction (or the exchange interaction, if there is one), which interchanges the spin orientation of two spins without changing the net magnetization: $I_i^+ I_j^- + I_i^- I_j^+$. It is important to realize that the process of spin diffusion involves no net change in Zeeman energy but only a spatial redistribution of it, if it is initially inhomogeneous, via a succession of mutual spin flips.

The rather specific analogy of the process to ordinary diffusion can easily be seen if we make a one-dimensional calculation. Consider a line of identical spin $\frac{1}{2}$ nuclei at x, $x \pm a$, $x \pm 2a$, and so forth. Let $P_{1/2}(x)$ and $P_{-1/2}(x)$ be the probabilities that the spin at x is in the $+\frac{1}{2}$ or $-\frac{1}{2}$ state. Define $p = P_{1/2} - P_{-1/2}$ (p is obviously the variable proportional to the magnetic moment density). Let the spins be coupled to nearest neighbors only by an exchange interaction: for the spin at x, the interaction Hamiltonian is

$$\mathscr{H}_{\text{int}} = \hbar J\{\mathbf{I}(x) \cdot [\mathbf{I}(x + a) + \mathbf{I}(x - a)]\} \tag{4-12}$$

The probability per unit time that a spin flips with a neighbor is on the order of J. Thus we can estimate

$$\frac{dP_{1/2}}{dt} = J[P_{1/2}(x + a) + P_{1/2}(x - a)]P_{-1/2}(x) - J[P_{-1/2}(x + a) + P_{-1/2}(x - a)]P_{1/2}(x) \tag{4-13}$$

If we use the relations $P_{1/2} + P_{-1/2} = 1$ and $P_{1/2} - P_{-1/2} = p$, and from them, $P_{+1/2} = (\frac{1}{2})(1 + p)$, $P_{-1/2} = (\frac{1}{2})(1 - p)$, we can rewrite Eq. (4-13), after a little algebra, in the form

$$\frac{dp}{dt} = \frac{J}{2}[p(x + a) + p(x - a) - 2p(x)] \tag{4-14}$$

The continuum equivalent to Eq. (4-14), expressed as a differential equation, is

$$\frac{dp}{dt} = D\frac{d^2p}{dx^2} \tag{4-15}$$

where $D = (\frac{1}{2})Ja^2$ is the spin diffusion constant. Equation (4-15) is the familiar diffusion equation. The one-dimensional model with nearest neighbor interactions does not do justice to the complexity of the three-dimensional problem with the longer range dipolar forces. The solution of the problem is by no means as straightforward as is the solution of the ordinary diffusion equation, but the analogy between ordinary diffusion and spin diffusion is the important thing to remember. For estimating the distance over which the quantity p changes in a given time, say T_1, one may still use as an estimate $(2DT_1)^{1/2}$. For dipolar coupling, $D \sim (1/30)(\Delta\nu^2)^{1/2}a^2$ provides a fair estimate [3]. The prefactor 1/30 is that small because the truncated dipolar Hamiltonian, from which $(\Delta\nu^2)$ is obtained, has other terms beside those of the form of Eq. (4-12), and has angular factors that must be averaged as well. If we set $a = 2 \times 10^{-8}$ cm, $(\Delta\nu^2)^{1/2} = 10^3$ Hz, then $D = 10^{-12}$ cm^2 sec^{-1}, and the diffusion length in a second ($T_1 \gg T_2$, then), is on the order of 10^{-6} cm, or 50 lattice constants. Thus, the event of a nuclear spin flipping over in a spin-lattice interaction is in a sense passed on to 10^4 other nuclei before the same spin flips again, on the average. In view of numbers such as these, it seems very fair to describe the spin system as a homogeneous thermodynamic system that interacts comparatively weakly with the outside world. For times much shorter than T_1, it can be regarded as an isolated system, and experiments done faster than T_1 are adiabatic in the thermodynamic sense.

One immediate consequence of these remarks is that the $T = T_S$ in Eq. (2-20) need not be the lattice temperature T_L but may be different from T_L. T_S is called the spin temperature. The spin temperature is not just defined as a number which, when plugged in place of T in Eq. (2-20), gives the Zeeman energy in field H_0. For the spin system to be in thermodynamic equilibrium and describable by T_S, it is required that the ratio of the spin populations of any pair of levels separated by ΔE be determined by the Boltzmann factor $\exp[-\Delta E/kT_s]$. Our argument about spin diffusion was to justify spatial homogeneity; that the dipole-dipole interaction can also bring about the correct population ratio from any non-equilibrium starting point is not so obvious, but a great deal of experimental experience has not turned up any evidence that it does not.

One method of changing the spin temperature of a spin system is to change the field rapidly compared to T_1 (but not so rapidly as to flip any spins). Since the only way the spin system can change its Zeeman energy is via the spin-lattice interaction, if we assume T_1 very long compared to the time scale of the change of H_0 from H_a to H_b, say, then Eq. (2-20) says that

$$T_b = T_a \frac{H_b^2}{H_a^2} \tag{4-16}$$

If H_a is large and H_b is small, Eq. (4-16) gives the final temperature T_b of an *adiabatic demagnetization experiment*, a very well known technique for obtaining very low temperatures with paramagnetic salts. One is tempted to set $H_b = 0$ and obtain $T_b = 0°\text{K}$. That is not possible because H_b^2 in Eq. (4-16) is, in fact, not just external field H_0^2 but is $H_0^2 + H_L^2$, where H_L is the "local field." H_L^2 is determined by the dipole interactions and is related to the second moment by $\gamma^2 H_L^2 = \frac{5}{3}\langle \Delta\omega^2 \rangle$.

There is no reason why the upper energy levels of the Zeeman ladder of Fig. 4-1 could not be more highly populated than the lower levels, and no reason either why the ratios of populations of different levels cannot be determined by the Boltzmann ratio. Under these circumstances, the temperature T_S would be negative. Negative absolute temperature is possible only if the system has the property that the highest energy level of the system is bounded. Clearly excluded by this requirement are systems involving kinetic energy, since in such systems the total energy of a system at negative absolute temperature would be infinite.

A spin system at negative spin temperature has the following properties just by virtue of $T_s < 0$. (1) It is not in equilibrium with the lattice, since the temperature for the lattice is necessarily positive. (2) It is "hotter" than any system of positive spin temperature in the sense that its Zeeman energy, given by Eq. (2-20), is greater than the energy of any system for which $T_s > 0$. (3) The "hottest" negative temperature is $T_s = 0-$; that is, the highest energy state of the system is the one for which all spins are in the highest level. Clearly a less confusing and more natural way to measure temperature is by the reciprocal of the absolute temperature, say $\beta_S = -1/T_S$. Then the hottest system has $\beta_S = +\infty$, and the coldest has $\beta_S = -\infty$.

A state of negative spin temperature may be obtained in a variety of ways. The simplest is to start with a spin system at the lattice temperature $T_L = T_S$ in field H_0. Apply a 180° rf pulse, which simply reverses the magnetization relative to H_0. Then $T_S = -T_L$. The first experiment demonstrating negative spin temperature was somewhat more elaborate [4]. The system was first demagnetized from the high field H_0, in which the resonance was to be observed, to a low field $H' > H_L$. The spin temperature is now low, given by Eq. (4-16). The field H' is the fringing field of the magnet that produced H_0, but at the position of H' there was also a solenoid. After the sample was placed in the solenoid, a field of approximately $2H'$ was produced in the solenoid in a direction opposite to the direction of H'. In this way, the field seen by the sample was reversed in approximately 0.2 μsec. The magnetization of the sample remains in its original direction, since the change in the direction of H' occurred quickly with respect to the natural time scale of the system, which is characterized by the Larmor frequency $\gamma H'$. We chose H' so that $2\pi/\gamma H'$ was longer than the field reversal time. Since M_0 is not antiparallel to H', the nuclear spin system is in a state of negative spin temperature: $T_S = -T_L H'^2/H_0{}^2$. Following the reversal of H', the current in the solenoid decayed to zero in a few milliseconds. That process is slow in the quantum mechanical sense, since it takes place in a time long compared to $1/\gamma H'$, but is adiabatic in the thermodynamic sense. Hence, the spin system maintains its negative spin temperature. The sample can now be reinserted into the high field, where its spin temperature is now $-T_L$.

The process is described graphically in Fig. 4-5, where the nuclear magnetization versus applied field is shown for the various steps. The sudden reversal of the field, step (b) in the figure, is shown dotted, since it is not a reversible thermodynamic process. The spin temperature corresponding to each stage of the experiment is also shown in Fig. 4-5. If the spin system behaved as we expected, the final nuclear magnetization will be antiparallel to H_0, and a nuclear resonance signal will correspond to stimulated emission rather than absorption. The magnetization will recover to its equilibrium value at the rate determined by the spin-lattice relaxation time. The signal recovery as a function of time is shown in Fig. 4-6.

We have not described the Pound–Purcell experiment because of its historical interest—it has been said that an active science tends to swallow its history—not because their method represents a particularly good way of achieving a negative spin temperature (it does not) but because it illustrates some steps and procedures characteristic of the manipulations one can perform on a spin system. These steps are the adiabatic ones in the thermodynamic sense of steps (a), (c), and (d) of Fig. 4-5, in which the intensive variable H is changed but there is no exchange of energy with a thermal reservoir. Then there is the sudden change of step (b). Step (b) is sudden in the quantum mechanical sense of the word in that the Hamiltonian is changed much more rapidly than the characteristic interaction time between the spin system and the external field.

The step (c) in Fig. 4-5 can be done in the rotating frame in such a way as to make unnecessary the steps (a), (b), and (d). The method of adiabatic fast passage was introduced by Bloch in his initial experiments on liquids. The technique as applied to solids (which, please remember, do not obey the Bloch equations) will require some further discussion, but

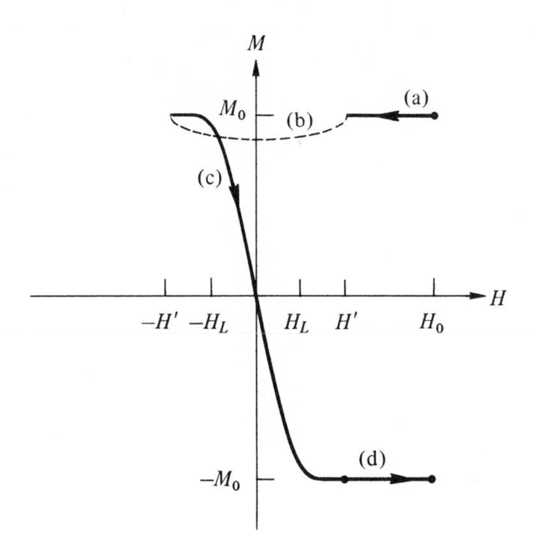

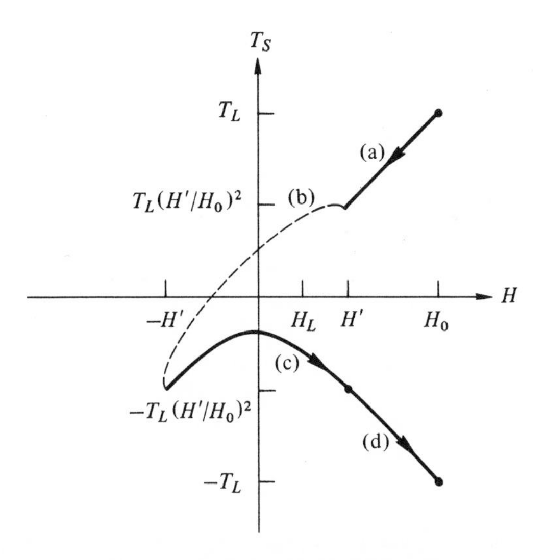

Fig. 4-5 Top: Sequence of nuclear magnetization changes in Purcell–Pound experiment [4]. Bottom: Corresponding sequence of nuclear spin temperatures. (a) Adiabatic demagnetization of $H' > H_L$. (b) Sudden reversal of field to $-H'$. (c) Adiabatic reversal of field to $+H'$. (d) Adiabatic magnetization to H_0. Nonadiabatic processes are shown as dashed lines to indicate that the path is meaningless on these diagrams.

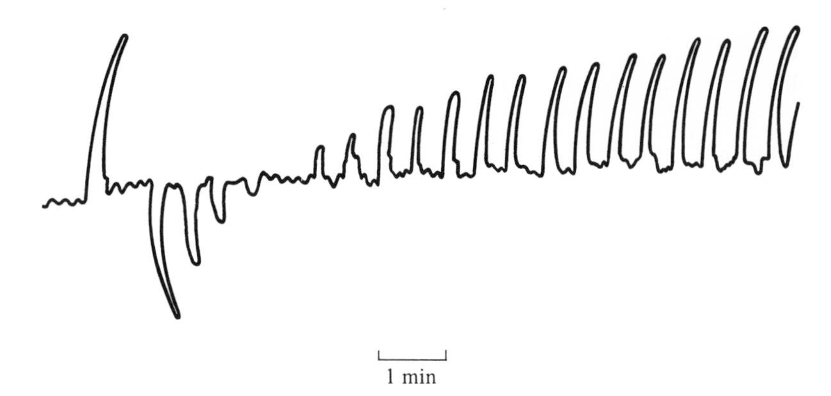

Fig. 4-6 ^{19}F nuclear resonance signal as a function of time subsequent to the establishment of a negative spin temperature in the ^{19}F system by the process of Fig. 4-5 (from reference [4]).

we shall describe the experiment and explain its justification afterward. Figure 4-7 represents a sequence of events as seen from the frame rotating at the frequency ω of a resonance experiment. A strong rf field is applied at ω, with, initially, $\omega < \omega_0$. Figure 4-7a shows the fields and magnetization as seen from the ω_0 frame. We assume, for the first discussion, that $H_1 \gg H_L$, but we shall weaken that assumption in the end. Figure 4-7b shows how the effective field changes in the rotating frame as H_0 is decreased through the resonance condition at ω—that is, $\omega = \gamma H_0$—and finally stopped substantially below the resonance field. The figure also shows M_0 following H_{eff} in the rotating frame, a process that is allowed quantum mechanically as long as the maximum reorientation rate of the effective field is slow compared with the characteristic time involved. In this case,

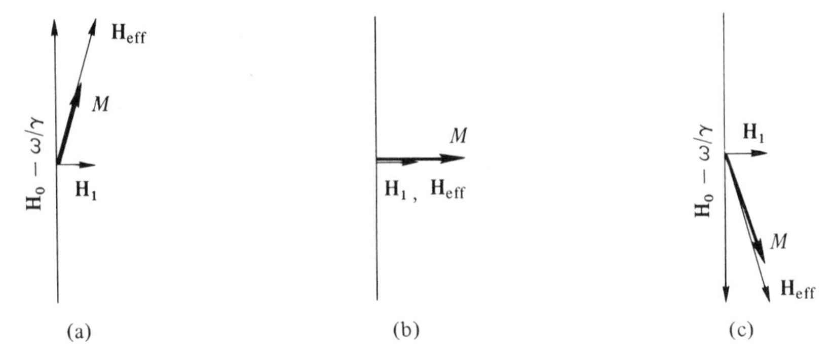

Fig. 4-7 Establishment of negative spin temperature by adiabatic rapid passage. Shown are effective fields in the rotating frame for H_0 above (a), equal to (b), and below (c) the field for resonance, ω/γ.

that time varies because H_{eff} changes magnitude as well as direction, unlike step (c) of Fig. 4-5. The necessary equivalent inequality is thus

$$\frac{dH_0}{dt} \ll \gamma H_1^{\ 2} \tag{4-17}$$

If the inequality Eq. (4-17) is satisfied, the state of affairs in Fig. 4-7c, where $|H_0 - \omega/\gamma| \gg H_1$, has M_0 pointing nearly antiparallel to H_0. If the whole process was carried through rapidly compared to T_1 (the "fast" in the name of the technique), nearly all the initial magnetization, initially pointing along H_0, is antiparallel to it, which is a state of negative spin temperature. One must ask, why does not M_0 decay in T_2 when it finds itself transverse to H_0, as in Fig. 4-7b? If H_1 is much larger than the local field, one can satisfy Eq. (4-17) and still travel through the "dangerous" region of substantial transverse magnetization in $t \ll T_2$. But in a solid, T_1 may be much longer than T_2, and we have already said that how quick "fast" must be is determined by T_1. In fact, we can still do the whole experiment as long as the resonance line is saturated, a condition described by the inequality $\gamma^2 H_1^{\ 2} T_1 T_2 \gg 1$, as suggested by the Bloch equations, where $T_2 \simeq 1/\gamma h_L$.

The process is explained by a hypothesis due to Redfield [5], which has been born out by numerous experiments and by some fundamental theoretical proofs. One begins by noting that if $H_1 \gg h_L$, a transverse reorientation of M_0 away from H_1 in Fig. 4-7b ought to proceed at the rate determined by T_1, not T_2, since it costs energy, γH_1 per spin, to reorient relative to H_1, and the characteristic dipolar energy per spin, γh_L, is much smaller than γH_1. This observation suggests that really the behavior of M_0 relative to H_{eff} in the rotating frame is much like its behavior in the Pound–Purcell experiment. So one should view M_0 as the magnetization of a system described by a Boltzmann distribution in the rotating frame with a spin temperature determined by analogy with Eq. (2-20) in the lab frame:

$$E_z^{rot} = -C\,\frac{(H_{eff}^2 + h_L^{\ 2})}{T_S^*} \tag{4-18}$$

The notation T_S^* is used to denote spin temperature in the rotating frame. If the analogy is correct, then the initial adiabatic fast passage experiment consists simply of a demagnetization down to the minimum value of $(H_{eff})_{min} = H_1$, and then a remagnetization with M_0 always pointing parallel to H_{eff}. Of course, H_{eff} ends up pointing antiparallel to its original direction, so M_0 does too, and the final state of affairs corresponds to a negative spin temperature as viewed from the laboratory frame. The use of the thermodynamic argument allows one to extend the discussion to $H_1 < H_L$, as long as H_1 is large enough to saturate the resonance. The case of $H_1 < H_L$ is just like adiabatic demagnetization in the laboratory frame to an external field comparable to, or less than, h_L. The order in the system, order that is manifested in high field by the magnetization of the sample, is transferred to the dipolar system, where the existence of order does not, however, produce a macroscopic polarization. (Remember, as the idea of order is suddenly thrust upon you, that the isolation from the lattice caused by the long T_1 means there is no heat flow to or from the spin system, so that the experiments are thermodynamically adiabatic, that is, isentropic.) As the sample is remagnetized, the order is put back into the Zeeman system, and the system remagnetizes along the same curve it demagnetized along. The process is reversible, and the full M_0 is recovered at the end. The only difference between the laboratory frame experiment of adiabatic demagnetization and the rotating frame experiment is that in the latter H_1 may not be reduced to zero, as H_0 may be in the laboratory frame, and the system still be remagnetized reversibly. Zero H_1, of course, means the rotating frame has no meaning, but it requires a rather more quantitative discussion to see that the requirement on H_1 is that $\gamma^2 H_1{}^2 T_1 T_2 \gg 1$ be satisfied.

We have discussed three ways to put a spin system in a negative spin temperature state: the 180° pulse, the sudden reversal of field method of Pound and Purcell, and the adiabatic fast passage technique. The real object of the discussion was not to convince you to master these techniques for themselves, but to convince you that in a solid with $T_1 \gg T_2$ the experimenter can manipulate the spin system with considerable freedom; he can prepare the system into a wide range of thermodynamic states. The practical consequences of these possibilities are, in fact, enormous. They range from the practical construction of the maser to the detection of the resonance of very rare (nonabundant) spins in a sample. We shall not attempt to describe any such experiments but take the discussion one step further and discuss what can be done when two spin systems are present in a sample.

If there are two spin species in a sample, and if their gyromagnetic ratios, γ_L and γ_S, are sufficiently different, they are almost totally independent thermodynamic systems. The requirement on the difference in the γ's is given by

$$|\gamma_I - \gamma_S|\, H_0 \gg \langle \Delta\omega^2 \rangle^{1/2} \tag{4-19}$$

where H_0 is the applied external field and $\langle \Delta\omega^2 \rangle$ is the larger of the second moments of the two resonance lines. Experimentally, Eq. (4-19) means that the lines do not overlap—they are clearly separated. The systems are independent; they are "insulated" from each other when Eq. (4-19) is satisfied because the exchange of Zeeman energy between the systems is energetically very unfavorable. Exchange of Zeeman energy could happen through the term in the dipole interaction of the form $I_i^+ S_j^-$, which causes an interchange of z component of the ith spin in the I system with the jth spin in the S system. The process causes a net change of Zeeman energy of the amount $\hbar\,|\gamma_I - \gamma_S|\,H_0$. If the inequality Eq. (4-19) is not satisfied, that unbalance in the Zeeman energy can be absorbed in the dipole-dipole energy of the systems by the rearrangement in their local fields of only a few spins. If Eq. (4-19) is satisfied, the dipole systems must absorb a large amount of energy, and that would require a large number of adjustments in local fields involving large numbers of spins, a very improbable process. So the important consequence of this chain of argument is that one may prepare two spin systems in the same sample at different spin temperatures. They will not exchange Zeeman energy as long as Eq. (4-19) is satisfied; they behave as if they are totally insulated from each other. There is, of course, the mutual contribution of the systems to the local field from the z component terms $I_i^z S_j^z$ in the dipole-dipole interaction, but that term causes no exchange of Zeeman energy between the systems.

If the inequality Eq. (4-19) is *not* satisfied, the spin systems are not totally independent. Clearly, Eq. (4-19) may be satisfied at sufficiently large fields. If the systems are prepared at very high field with, say, $T_S = \infty$ for the I system, and $T_S = T_L$ for the S system, and the field is then lowered until the resonance lines begin to overlap, mutual spin flips will cause the Zeeman energy to flow from the I system to the S system, much as if a heat switch had been closed between them. Schematically, the situation can be represented by Fig. 4-8. The time T_{cr} which describes

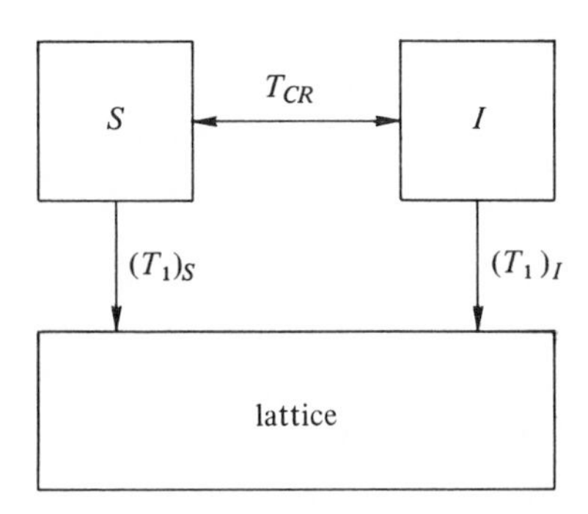

Fig. 4-8 Spin systems S and I at spin temperatures T_S and T_I, weakly coupled to each other with cross-relaxation time T_{cr}, and to the lattice with spin-lattice relaxation times $(T_1)_S$ and $(T_1)_I$, respectively.

the time scale of interaction between the systems is known as the *cross-relaxation time*. One can see immediately some of the wealth of possibilities inherent in the situation. For example, if $(T_1)_S \gg T_{cr}$ and $(T_1)_S \gg (T_1)_I$, the S-system relaxation is determined by T_{cr}, which is not a spin-lattice relaxation process at all. Or, it may be that $T_{cr} \ll (T_1)_I$, in which case the two spin systems equilibrate with each other and relax to the lattice at the common rate determined by $(T_1)_I$. It is common to refer to the slower of the relaxation rates in a process in which the mechanisms are in series, or sequential, as a "bottleneck." In the first of the preceding suppositions, the cross-relaxation process was the bottleneck. In the second, the $(T_1)_I$ process formed the bottleneck.

The ability to prepare materials with appropriate cross-relaxation rates was very important in the development of the maser. The full range of possibilities cannot be explored, however, if one is confined to spin systems with equally spaced energy levels in a magnetic field. We shall not explore the complications of cross relaxation further, but shall turn to the quadrupole interaction between nonspherical nuclei and electric field gradients because of its great importance to nuclear magnetic resonance.

4-4. NUCLEAR QUADRUPOLE INTERACTION

Most elementary courses in electricity and magnetism introduce the multipole description of charge distribution by considering the special cases of arrangements of point charges. Figure 4-9 summarizes these standard arrangements of point charges that possess only a 2^l-pole moment, where $l = 0, 1, 2, \ldots$ (monopole, dipole, quadrupole, ...).

Nuclei are not point charges, but may for our purposes be regarded as geometrical shapes with at least a monopole moment $+Ze$. The monopole moment can then be written in terms of the charge density $\rho(\mathbf{r})$:

$$Ze = \int_V \rho(\mathbf{r})\, dV \tag{4-20}$$

The integral is over all space, but $\rho(\mathbf{r})$ is always well localized near $\mathbf{r} = 0$ and drops off rapidly beyond a certain radius. The dipole moment in the

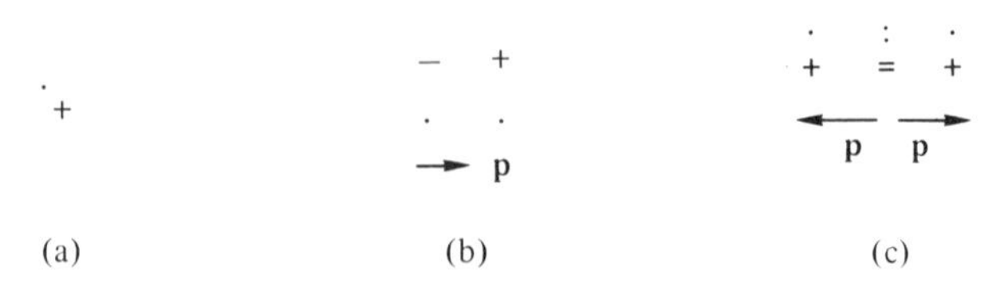

Fig. 4-9 Point charge distributions that have only an electric monopole (a), dipole (b), and quadrupole (c).

presence of a charge density function (rather than just a point charge as in Fig. 4-9) is the vector **p** defined by

$$\mathbf{p} = \int_V \rho(\mathbf{r})\mathbf{r}\, dV \tag{4-21}$$

If the charge distribution is spherical, $\rho(\mathbf{r}) = \rho(r)$. Now **r** is an odd function and $\rho(\mathbf{r})$ is an even function, so $\mathbf{p} = 0$. The statement that $\rho(\mathbf{r})$ is even, or $\rho(\mathbf{r}) = \rho(-\mathbf{r})$, is a nontrivial one. The quantum expression for $\rho(\mathbf{r})$ is

$$\rho(\mathbf{r}) = e\,|\psi(\mathbf{r})|^2 \tag{4-22}$$

where $\psi(\mathbf{r})$ is the nuclear wavefunction, in general a many-body wavefunction involving all the protons in the nucleus. If $\psi(\mathbf{r})$ is either an even or an odd function—that is, if $\psi(\mathbf{r})$ has a definite *parity*—then $\rho(\mathbf{r})$ is even. Since we are dealing with the ground state, which, from experimental evidence on all known nuclei, is a nondegenerate state, it is a state of definite parity. Put another way, were the ground state doubly degenerate, and the two wavefunctions of opposite parity, then the ground state wavefunctions would be linear combinations of even and odd functions, and $\rho(\mathbf{r})$ would have an odd component. Such is never the case, and the electric dipole moment of the ground state of all nuclei vanishes.

The inexorable increase of complication suggests that the next multipole moment will be a tensor quantity, since the monopole moment was a scalar, the dipole a vector. A tensor of the second rank is in general a nine-component quantity, but the electric quadrupole moment is a traceless symmetric tensor, with only five independent components:

$$Q_{\alpha\beta} = \int [(3X_\alpha X_\beta - \delta_{\alpha\beta} r^2)\rho(r)]\, d^3r \tag{4-23}$$

where $\alpha, \beta = x, y, z$. The second term in the integrand was introduced because, as defined by Eq. (4-23), the quadrupole term will depend only on the orientation of the nucleus. Equation (4-23) defines a traceless quantity: $\sum_\alpha Q_{\alpha\alpha} = 0$. The traceless property gives a relation among the diagonal elements of the symmetric tensor (which has six components), leaving only five components necessary to specify the nuclear charge distribution.

It is best to remember the multipole properties of the nuclear charge distribution in terms of a geometrical shape. A nucleus with only a monopole moment has a spherical charge distribution. A nucleus with a quadrupole moment has a ellipsoidal charge distribution. More complete treatments of the problem than we present here show that if the nucleus is in a definite state of angular momentum it must have cylindrical symmetry, so the ellipsoid is one of revolution about an axis. This fact, combined with the tracelessness of Eq. (4-23), means that only one number (of the original possibility of nine) is needed to specify the nuclear quadrupole moment. The definition in terms of the charge distribution $\rho(\mathbf{r})$ is

$$eQ = \tfrac{1}{2} \int (3z^2 - r^2)\rho(\mathbf{r})\, d^3r \tag{4-24}$$

[Check that Eq. (4-24) vanishes if $\rho(\mathbf{r})$ is a function of $|\mathbf{r}|$ only.] Although it is now known that nuclei have more complicated shapes than just ellipsoids of revolution, and must be specified by higher multipole moments (octupole, hexadecapole, etc.), examples of the effects of higher moments on resonance experiments are rare.

Next we are interested in energies of interaction of the nuclear charge distribution with electric fields. The energy of a monopole depends on its *position* in a uniform electric field, that of a dipole on its *orientation*, but not on its position. By extrapolation, it follows that the energy of a quadrupole moment is independent of its *position and orientation* in a *uniform* electric field. The student should convince himself of these statements with the distribution of Fig. 4-9. Investigations with two-dimensional figures are quite helpful in cutting through a lot of mathematics. For example, the three parts of Fig. 4-10 show intuitively the orientation dependence of the energy of an elliptical charge distribution with the nonuniform electric field produced by a point charge. If the ellipse is made of positive charge, Fig. 4-10b is lower in energy than Fig. 4-10a. Figure 4-10c "differs" from 4-10a because it has been arrived at by rotating the ellipse 180° from (a). Of course, the ellipse has twofold

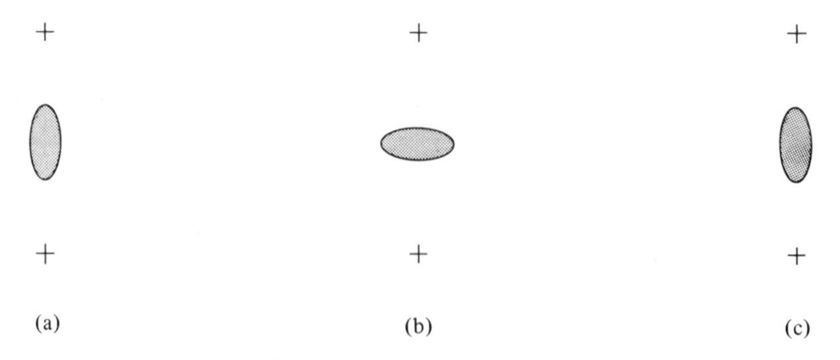

Fig. 4-10 The shaded ellipse, which represents a figure of revolution about the major axis, is a quadrupole moment in a field gradient created by the point charges. See text for explanation.

symmetry—it looks the same after a 180° rotation, so Fig. 4-10c is exactly the same as Fig. 4-10a, a result we shall find to be of central importance soon.

Further insight can be obtained by considering the charge distributions of Fig. 4-11. The four positive charges are arranged on the corners of a square, and the quadrupole in the *exact center* of the square has the same energy in parts (a), (b), and (c) of Fig. 4-11. In Fig. 4-11c, the angle θ is arbitrary. That the energies of parts (a) and (b) are the same is clear from symmetry. That the same is true for (c) is not so obvious, although it also follows from a deeper application of the idea of symmetry. The student should verify it for the "point" quadrupole of Fig. 4-9c—that is, the energy is independent of θ in the limit of a point quadrupole centered in the square.

The description of the shape of the nucleus in terms of multipole moments follows naturally from the expression for the energy of a charge distribution in an electric potential:

$$U = \int \rho(\mathbf{r}) V(\mathbf{r}) \, d^3r \tag{4-25}$$

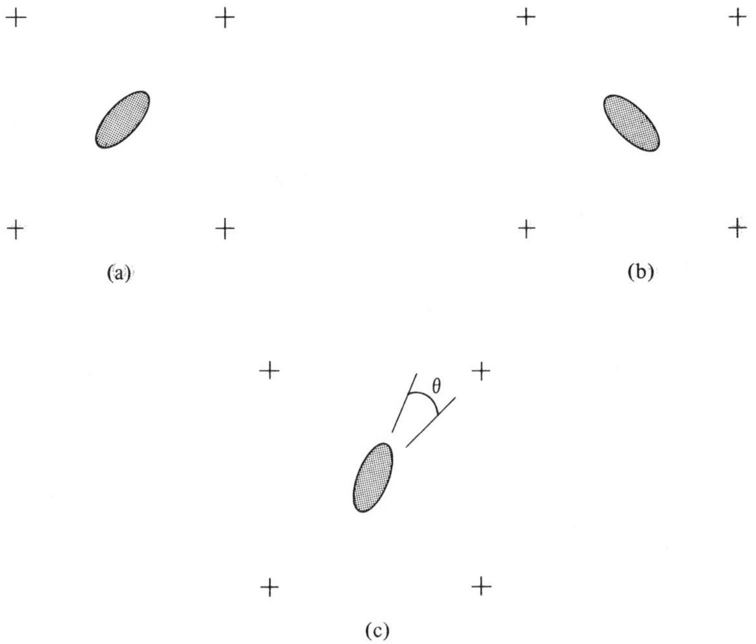

+ + + + + + + +

(a) (b)

+ + θ + +

(c)

Fig. 4-11 The interaction energy of the point quadrupole in the center of the square with point charges at the corners is the same in (a), (b), and (c).

If $\rho(\mathbf{r})$ is confined to a region near $\mathbf{r} = 0$, we can expand $V(\mathbf{r})$ in a Taylor's series about $\mathbf{r} = 0$:

$$V(\mathbf{r}) = V(0) + \sum_i u_i V_i(0) + \frac{1}{2!} \sum_{i,j} u_i u_j V_{ij}(0) + \cdots \tag{4-26}$$

where

$$V_i = \frac{\partial V}{\partial u_i} \qquad V_{ij} = \frac{\partial^2 V}{\partial u_i \, \partial u_j}$$

and

$$u_i = x, y, z$$

With these definitions, Eq. (4-25) becomes

$$U = V(0) + \sum_i V_i \int u_i \rho \, d^3 r + \frac{1}{2!} \sum_{ij} V_{ij} \int u_i u_j \rho \, d^3 r + \cdots \tag{4-27}$$

The first term of Eq. (4-27) is a constant, the electrostatic energy of the nucleus. The second term is the dipole term, which vanishes, as we argued previously, and the third term is the quadrupole term. In terms of Eq. (4-23),

$$\int u_i u_j \rho \, d^3 r = \frac{1}{3} \left(Q_{ij} + \int \delta_{ij} r^2 \rho \, d^3 r \right)$$

so the quadrupole term becomes

$$U^{(2)} = \frac{1}{6} \sum_{ij} V_{ij} \left(Q_{ij} + \int \delta_{ij} r^2 \rho(r) \, d^3 r \right) \tag{4-28}$$

If the charge distribution that produces the potential V vanishes at $\mathbf{r} = 0$, $V(\mathbf{r})$ obeys Laplace's equation at $\mathbf{r} = 0$, and the second term of Eq. (4-28) is

$$\sum_{ij} V_{ij} \delta_{ij} \int r^2 \rho \, d^3 r = \left(\sum_j V_{jj} \right) \int r^2 \rho \, d^3 r = 0$$

since $\sum_j V_{jj} = 0$ is Laplace's equation in this notation. If the charge distribution is *not* zero, Poisson's equation is obeyed:

$$\sum_j V_{jj} = -4\pi\rho(0) = 4\pi e \, |\psi(0)|^2$$

where $|\psi(0)|^2$ is the probability density of electrons at the nucleus. Now, the second term, although not zero, is independent of indices i, j; that is, it is orientation independent. The term is of consequence in the Mössbauer effect—it produces the so-called isomer shift—but it has no consequences in nuclear magnetic resonance.

There are now two steps remaining to be done to convert the classical expression for the electric quadrupole-electric field gradient interaction into an expression from which energy levels may be calculated. The expression we need to investigate is simply the first part of Eq. (4-28)

$$E^{(2)} = \tfrac{1}{6} \sum_{ij} V_{ij} Q_{ij} \tag{4-29}$$

The most important step concerns the quadrupole moment Q_{ij}, which we want to convert to a quantum mechanical operator. We refer the reader to Eq. (1-1), where the Wigner–Eckart theorem was quoted in justifying the relation between magnetic moment and angular momentum. The proportionality presented there exists because of the similarity between the transformation properties (i.e., behavior upon rotation) of the magnetic moment and the angular momentum. The next step in our present problem is exactly the same; as before, we shall not present the argument, just the results. A good discussion of the Wigner–Eckart theorem applied to the quadrupole operator is in Slichter [2]. The result is

$$Q_{ij} = \frac{eQ}{I(2I-1)} \left[\tfrac{3}{2}(I^i I^j + I^j I^i) - \delta_{ij} I^2\right] \tag{4-30}$$

where I in the prefactor is the angular momentum of the nucleus having quadrupole moment Q given by Eq. (4-24). The operator part of the equation, in brackets, involves quadratic forms of I, as one perhaps should expect. The quantum mechanical expectation value of I^2 is, of course, $I(I+1)$. Notice that Eq. (4-30) does not work for $I = \frac{1}{2}$ because of the $2I - 1$ in the denominator. The quadrupole moment Q of a spin $\frac{1}{2}$ nucleus is, in fact, zero; a spin $\frac{1}{2}$ nucleus has a spherically symmetric charge distribution (see Cohen and Reif [6], or Das and Hahn [7]).

Before we insert Eq. (4-30) into (4-29), we can simplify the expression we shall get by examining the matrix $\{V_{ij}\}$. We can use Laplace's equation $\sum_j V_{jj} = 0$, since, as we have already argued, a nonzero charge at $r = 0$ has spherical symmetry, so that it produces no effects depending on the orientation of the nucleus. Because V_{ij} is a second derivative of an analytic function, $V_{ij} = V_{ji}$. Thus, $\{V_{ij}\}$ has five independent elements: three to specify the orientation and just two that have to do with the magnitude and "shape" of the field gradient. The orientation may be specified by finding the coordinate system in which $\{V_{ij}\}$ is diagonal. The other two parameters conventionally used to specify $\{V_{ij}\}$ are then the magnitude eq, and the asymmetry parameter η:

$$eq = V_{zz}$$
$$\eta = \frac{V_{xx} - V_{yy}}{V_{zz}} \tag{4-31}$$

In Eq. (4-31), the xyz coordinate system is the principal axis system of $\{V_{ij}\}$, and the axes have been chosen such that $V_{zz} > V_{xx} > V_{yy}$, so that $0 < \eta < 1$. (Note that because $V_{xx} + V_{yy} + V_{zz} = 0$, with this ordering of the elements of $\{V_{ij}\}$, $V_{zz} > 0$ always.) The special case of $\eta = 0$, which is often encountered, is known as *axial symmetry*. With these final definitions, we may now write the quadrupole Hamiltonian:

$$\mathscr{H}_Q = \frac{e^2qQ}{4I(2I-1)}[(3I_z^{\,2} - I^2) + \eta(I_x^{\,2} - I_y^{\,2})] \tag{4-32}$$

To Eq. (4-32) must be added the Zeeman interaction, $\mathscr{H}_Z$, of the nuclear dipole moment with an external magnetic field. There are two limiting cases, $\mathscr{H}_Z \ll \mathscr{H}_Q$, and $\mathscr{H}_Z \gg \mathscr{H}_Q$, for which the energy levels may be obtained to excellent accuracy using the standard approximation techniques of quantum mechanics. We shall sketch the procedure and discuss the results for both limits for an axial field gradient, $\eta = 0$.

When $\mathscr{H}_Z = 0$, Eq. (4-32) still produces a splitting between energy levels, and transitions between the levels may be observed by almost all the standard magnetic resonance techniques. The energy levels of Eq. (4-32), with $\eta = 0$, may be found immediately and exactly by computing $\langle mI|\mathscr{H}_Q|mI\rangle$:

$$E(m) = \frac{e^2qQ}{4I(2I-1)}[3m^2 - I(I+1)] \tag{4-33}$$

Note that the ellipsoidal shape of the nucleus is reflected in Eq. (4-33) by the degeneracy of $E(m)$ and $E(-m)$. That is, it does not matter which way the nucleus points—it is, as far as we can tell here, invariant upon rotation by 180° about any axis perpendicular to its axis of rotation. The energy levels of Eq. (4-33) for $I = 1, \frac{3}{2}, 2, \frac{5}{2}$ are shown in Fig. 4-12, where we have chosen $e^2qQ \propto 4I(2I-1)$ as a scale factor. There are several interesting aspects of Eq. (4-33) and Fig. 4-12. The odd half-integral spins are split into doublets; the integral spins always have the singlet $m = 0$ state lowest (for $Q > 0$, of course). The inability of the electric quadrupole inter-

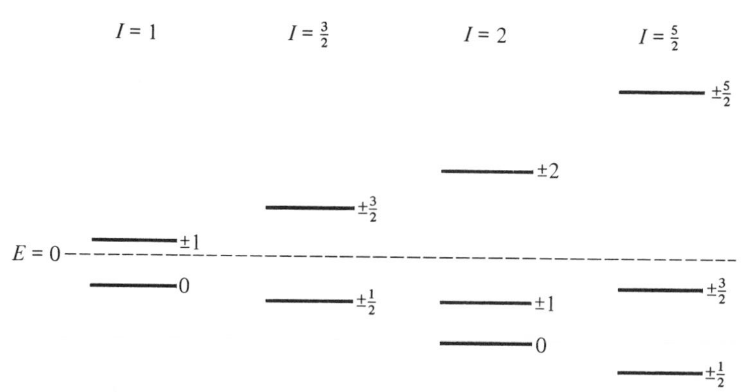

Fig. 4-12 Energy-level diagram of Eq. (4-33) for spins $I = 1$, $\frac{3}{2}$, 2, and $\frac{5}{2}$. The scale has been determined by letting e^2qQ be proportional to $4I(2I-1)$, and $Q > 0$.

action to remove the two fold degeneracy for the odd half-integral spins is an example of a theorem due to Kramers that deals with the effect of electric fields on systems with an odd number of spin $\frac{1}{2}$ particles, or, equivalently, with systems the total angular momentum of which is half-integral. The twofold degeneracy which that theorem says must remain in the presence of just electrostatic interactions is known as *Kramers' degeneracy*. The proof of the theorem depends on quite general properties of the system under time reversal.

Transitions between adjacent energy levels of Fig. 4-12 may be induced by an rf magnetic field at the appropriate resonance frequencies in the direction transverse to the axial field gradient direction. Such an experiment is known as pure quadrupole resonance. If the field is linearly polarized, say along the x direction, then one may regard one of the rotating components as producing transitions between the m and $m-1$ components of adjacent levels, with the induced magnetization circularly polarized in the same sense. The counterrotating component of the applied field produces transitions between the $-m$ and $-m+1$ components of the same pair of levels. The total induced magnetization is the sum of two equal circularly polarized magnetizations, and is consequently linearly polarized parallel to the applied rf field. (Refer to the remark on p. 48 about observation of pure quadrupole resonance with a crossed-coil apparatus. If $\eta > 0$, the induced magnetization is, in general, elliptically polarized, and $\chi_{xy} \neq 0$.)

A small magnetic field splits the degenerate levels of Fig. 4-12. If H_0 is applied in the z direction, the problem may be solved almost by inspection, since the Zeeman perturbation is diagonal in the coordinate system of the axial electric field gradient. If H_0 is applied in some other direction, the problem is considerably more complicated, and the splitting depends on the angle between H_0 and z. If $\eta \neq 0$, the angular dependence of the splitting also depends on η, and the study of Zeeman splittings of the pure quadrupole resonance provides the standard way of measuring η. Figure 4-13 shows the Zeeman splittings and resonance transitions seen for the case of $I = 1$ and $I = \frac{3}{2}$. Note that the frequency of the spectrometer is swept in order to present resonance curves of the type shown in Fig. 4-13. Sweeping frequency rather than field is necessarily the standard technique in pure quadrupole resonance.

In the other limit, when $\mathscr{H}_Z \gg \mathscr{H}_Q$, the quadrupole interaction is treated as a small perturbation on the otherwise equally spaced energy levels of the Zeeman interaction. It is again easy to see what happens if H_0 is applied parallel to the electric field gradient axis for $\eta = 0$. The Hamiltonian is

$$\mathscr{H} = \mathscr{H}_Z + \mathscr{H}_Q = -\gamma\hbar H_0 I_z + \frac{e^2qQ}{4I(2I-1)}[3I_z{}^2 - I^2] \tag{4-34}$$

Since Eq. (4-34) involves only I_z, it is diagonal in the m representation, and the exact energies may be written down:

$$E(m) = -\hbar\omega_0 m + \frac{e^2qQ}{4I(2I-1)}[3m^2 - I(I+1)] \tag{4-35}$$

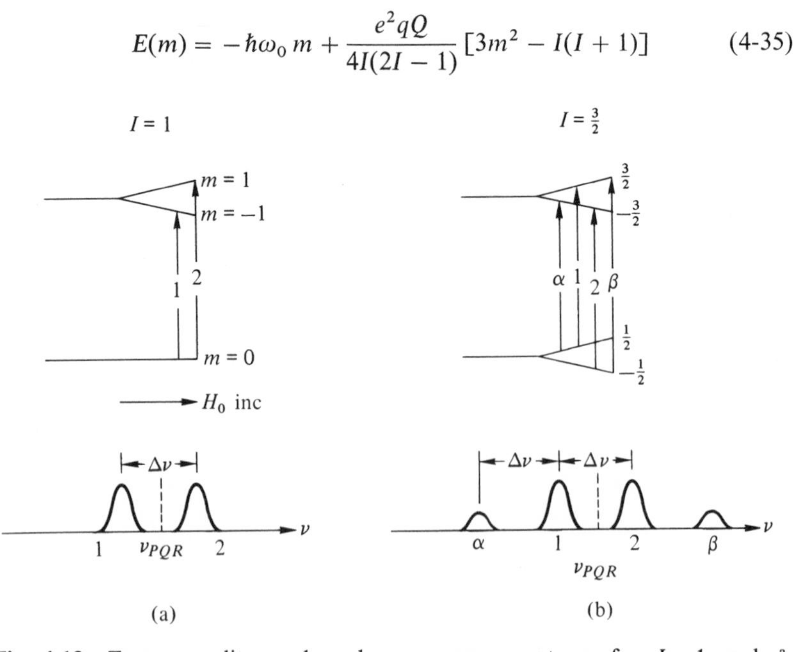

Fig. 4-13 Zeeman split quadrupole resonance spectrum for $I = 1$ and $\frac{3}{2}$. If H_0 is parallel to the z axis of an axial electric field gradient, then $\Delta\nu = (1/2\pi)\gamma H_0$. The α, β transitions occur if $\eta \neq 0$ or if H_0 is not parallel to the field gradient axis.

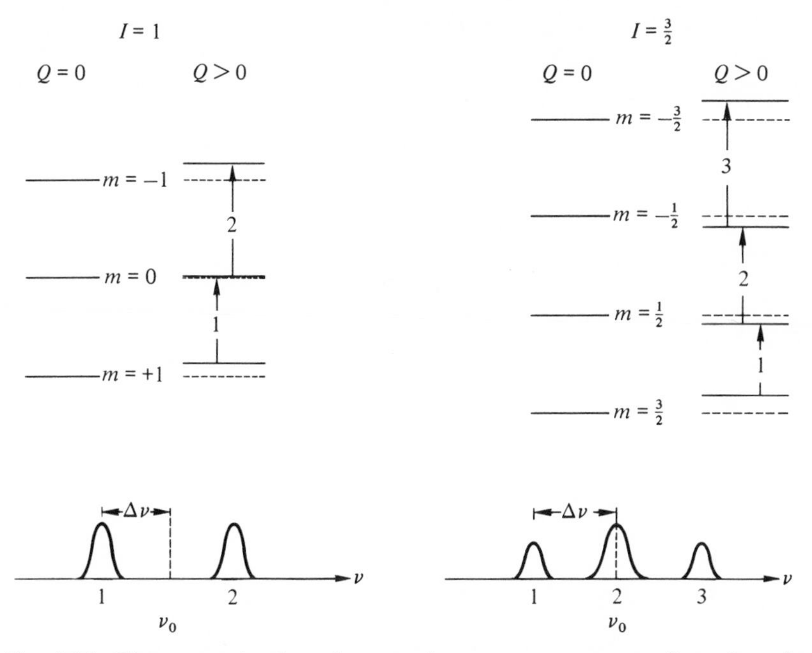

Fig. 4-14 The energy levels and magnetic resonance spectra of quadrupoles of $I = 1$ and $\frac{3}{2}$ in a large magnetic field and an axial electric field gradient. The splitting $\Delta\nu$ is given by

$$\Delta\nu = \frac{3e^2qQ}{4I(2I-1)}(2m-1)$$

Figure 4-14 shows the energy levels of Eq. (4-35) for $Q = 0$, and then for $Q > 0$, for $I = 1$ and $\frac{3}{2}$. The characteristic features of Fig. 4-14 are as follows. For either integer or half odd-integer spin the $2I$ degenerate pure Zeeman transitions at ν_0 are split into $2I$ lines. For I a whole integer, the splitting is symmetric about ν_0, with, since $2I$ is even, no transition at ν_0. For $2I$ odd, the $m = \frac{1}{2}$ to $m = -\frac{1}{2}$ transition remains at ν_0 (the $\pm\frac{1}{2}$ levels are lowered an equal amount by the quadrupole interaction), and the other transitions, known as "satellites," are split symmetrically from ν_0. The spectrum for $I = \frac{3}{2}$ in Fig. 4-14 has been drawn to indicate that the ratios of the intensities of the three lines are as 3 : 4 : 3. This result simply comes from the fact that one of the quantities that determine the intensity of the magnetic resonance absorption, $|\langle m| I_x |m-1\rangle|^2$, is in the ratio 3 : 4 : 3 for $m = \frac{3}{2}, \frac{1}{2}$, and $-\frac{1}{2}$. For $I = \frac{3}{2}$, then, 40% of the intensity of the resonance transition comes from the central line.

If H_0 is applied at an angle θ with respect to the principal axis of the electric field gradient, then the splitting $\Delta\nu$ depends on angle. In first order, the problem is not difficult to solve. One simply writes $\mathscr{H}_Q$ in the Zeeman system, using the relation

$$I_z = I_{z'} \cos\theta + I_{x'} \sin\theta \tag{4-36}$$

Figure 4-15 shows the relation between the primed coordinate system, with H_0 parallel to z', and the electric field gradient principal axis system, with z the axis when $\eta = 0$. When the I_z of Eq. (4-36) is squared and substituted into Eq. (4-34), the first-order approximation to the energy levels may be found without too much labor (see Slichter [2], p. 172). The result is

$$E(m) = -\hbar\omega_0 m + \frac{e^2qQ}{4I(2I-1)} \frac{(3\cos^2\theta - 1)}{2} [3m^2 - I(I+1)] \tag{4-37}$$

The splitting of each satellite line is a function of field orientation with the characteristic $(3\cos^2\theta - 1)$ pattern. Note all lines coalesce to this order when $\cos^2\theta = \frac{1}{3}$, or $\theta \simeq \pm 55°$.

We remind the reader that the results quoted are for $\eta = 0$. They are also lowest-order corrections to the Zeeman energy. The effect of the quadrupole interaction of shifting both $m = \frac{1}{2}$ and $-\frac{1}{2}$ levels equally, and hence leaving the transition between them unchanged in frequency from γH_0, is strictly a first-order effect. In second order, and certainly in an

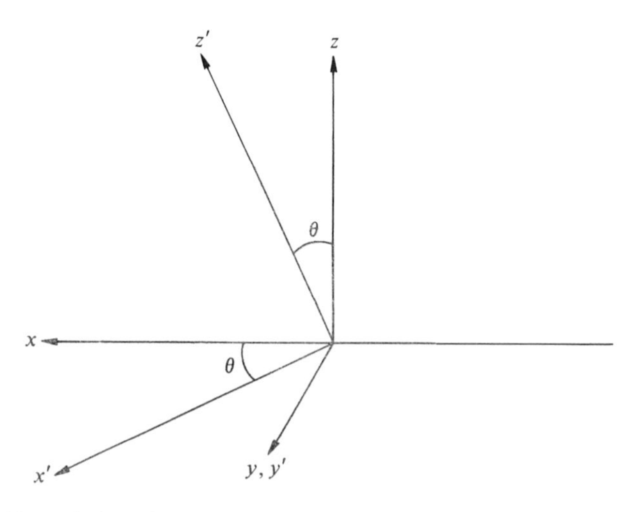

Fig. 4-15 Relation between the primed coordinate system (x', y', z'), with H_0 parallel to z' and the unprimed principal axis system of the electric field gradient.

exact calculation, there is some angular dependence of the frequency of the central transition.

The preceding pages have described the electric quadrupole interaction. It is, regrettably, algebraically tedious to deal with. What physics is there to it? The answer is to examine the origin of electric field gradients. In a free atom or molecule, the electric field gradient, if there is one, is caused by the electrons on the atom or molecule. A problem at the end of the chapter leads you through the computation of the electric field gradient at the origin because of an electron at (x, y, z). If the electron is in an atomic p_z state (to pick the simplest example), one can show that the electric field gradient at the origin is $-(4/15)|e|\langle 1/r^3 \rangle$, where $\langle 1/r^3 \rangle$ is an average over the radial wavefunction of the p_z electron. If the problem were entirely a one-electron one, then for a single atom at least the discussion would be over with that remark. But the $\langle 1/r^3 \rangle$ suggests strongly that electrons near the nucleus are the most important. If our p electron is a single electron outside a closed (spherically symmetric) shell, or a hole in an otherwise filled p shell (such as in a halogen atom), the small effects of the interaction of the valence electron with the closed shell electrons can have a very large effect on the field gradient experienced by the nucleus through distortion of the spherical symmetry of the close-in core electrons. Rather difficult calculations of the effect have shown enhancements of the electric field gradient from a distant charge by distortion of the closed shell as much as 100 times. (The enhancement is called the *Sternheimer antishielding factor.*) The need to rely on such calculations to determine the true q seen by a nucleus is a major obstacle to determining with precision the quadrupole moment Q of a nucleus. Fortunately, the systematic study of the hyperfine interaction has enabled some Q values to be reliably determined, and gradually meaningful values for most nuclei have been accumulated.

In a solid, an electric field gradient at a nucleus may be produced by the combination of covalent bonding to nearest neighbors (which involves p electrons), the sum of ionic charges over the entire lattice, and/or the charge distribution of conduction electrons in a noncubic metal. The systematic study of electric quadrupole interactions has been extremely useful in helping to determine the nature of the binding in many solids.

In addition, time dependent quadrupole interactions are an effective relaxation mechanism. We shall discuss in greater detail some relaxation mechanisms in solids. For completeness, we mention here that fluctuating electric field gradients in liquids are often important contributors to T_1. An example will serve to remind us of an effect discussed in Chapter 3. Both isotopes of bromine have large quadrupole moments and bromine has a large antishielding factor. In BrF_5, four of the five fluorines are equivalent. Hence, the spin-spin coupling among the fluorines splits the fluroine nuclear resonance into a quintet of total relative intensity of one, and a doublet, total relative intensity four. Although ^{79}Br and ^{81}Br both have $I = \frac{3}{2}$, there is no evidence of coupling seen between the Br and F nuclei in the spectrum. The explanation, as given in Chapter 3, is to be found in the short T_1 of Br, which "washes out" the coupling (modulation index much less than one). The Br T_1 is short because the Br nuclei relax via a time dependent quadrupole interaction in the liquid. The ^{19}F spins are $I = \frac{1}{2}$, of course, and so they relax by the slower magnetic dipole-dipole mechanism. The fluctuating field gradients in a liquid can be caused by distortions of the electron charge clouds on the molecule during collisions, or by free ions in the liquid.

4-5. SPIN-LATTICE RELAXATION IN SOLIDS

A complete survey of spin-lattice relaxation in solids would properly occupy a fair-sized book, particularly if it included the relaxation of electronic spins in paramagnetic and ordered magnetic systems. In this section we shall discuss only the relaxation of nuclei in diamagnetic insulators, with emphasis on the interaction between nuclei and lattice vibrations. Metals will be discussed in Chapter 5.

Nuclear spins have either spin $\frac{1}{2}$, without a quadrupole moment, or they have, with few exceptions, an appreciable quadrupole moment. If a nucleus does not have a quadrupole moment, its spin-lattice relaxation rate in a solid far from the melting temperature is slow and depends on the purity of the solid. That its relaxation rate should be slow will soon become apparent. We shall discuss first the usual relaxation mechanism for spin $\frac{1}{2}$ nuclei in a rigid lattice. The mechanism was first investigated by Bloembergen in the paper referred to in connection with spin diffusion [3].

An electronic spin attached to a paramagnetic ion is much more tightly coupled to the lattice than a nuclear spin, only in part because of the much larger magnetic moment. The relaxation time at room temperature for the spin of a paramagnetic impurity can range from the millisecond region down to less than 10^{-10} sec. A nucleus close to a paramagnetic impurity sees a fluctuating magnetic field with a correlation function given by

$$f(\tau) = \langle H(t)H(t+\tau)\rangle = \langle H(t)^2\rangle \exp \frac{-\tau}{T_{1S}} \tag{4-38}$$

where T_{1S} is the paramagnetic spin relaxation time. The mean square field, $\langle H^2\rangle$, is on the order of $g\beta_0 S_z/r^3$, where $g\beta_0 S_z$ is the magnetic moment of the paramagnetic impurity and r the distance from the nucleus to the paramagnetic impurity. We saw in Chapter 3 that the nuclear T_1

is proportional to the Fourier transform of Eq. (4-38) at the nuclear Larmor frequency, $\omega_0 = \gamma_n H_0$ [Eq. (3-29)]. Now a nucleus can be in three relatively distinct regions relative to the paramagnetic impurity. It can be so far away that its relaxation time due to its direct coupling with the impurity can be neglected (region I). It can be so close that not only is $\langle H(t)^2 \rangle$ large, but its z component is larger than local field from nuclear dipole interactions (region III). Under these circumstances, incidentally, the appropriate z component might be proportional to either $g\beta_0 S_z$ or $g\beta_0 \langle S_z \rangle$, where $\langle S_z \rangle$ is the thermal average of S_z:

$$\langle S_z \rangle = \frac{g\beta_0 S(S+1)}{3kT} H_0 \tag{4-39}$$

Equation (4-39) follows from the expression for the paramagnetic susceptibility of a spin system [Eq. (2-20)]. The choice between S_z or the greatly reduced $\langle S_z \rangle$ depends on T_{1S}. One uses S_z if T_{1S} is long enough that the nuclear precession in the field of the impurity can follow the change in the impurity's spin orientation—that is, if $\gamma_n[\langle H(t)^2 \rangle]^{1/2} T_{1S} \gg 1$. If $\gamma_n[\langle H(t)^2 \rangle]^{1/2} T_{1S} \ll 1$, then region III nuclei are sensitive only to $\langle S_z \rangle$, the thermal average of S_z. In region III, which is close to the impurity, it is usually the case that the z component of the field produced by the impurity varies so rapidly ($\sim 1/r^3$) that the Larmor frequency of nuclei from shell to shell as one goes away from the impurity differs by more than the nuclear dipole fields. Hence, nuclear spin flips between pairs of nuclei in different shells do not conserve energy.

Region II nuclei are those nuclei between regions I and III. Their relaxation time is determined by the impurity, but their Larmor frequency is $\omega_0 = \gamma_n H_0$. Region II nuclei relax rapidly; they are in good contact with the lattice. Region I nuclei sense the temperature of region II nuclei by spin diffusion, the temperature independent mechanism we discussed at the beginning of this chapter. Region III nuclei are also tightly coupled to the lattice by their strong interaction with the paramagnetic spins, but are otherwise not really part of the nuclear spin system, since their Larmor frequencies are, by definition, different. The regions are illustrated in Fig. 4-16.

Suppose one saturates the nuclear resonance with an rf pulse and watches the recovery of the spin temperature. The region III spins are unlikely to be affected, unless one takes special pains. Region II nuclei are saturated, but they recover quickly to the lattice temperature. Region I nuclei reach the lattice temperature only via the very slow process of spin diffusion. The nuclear resonance signal seen during recovery contains contributions from nuclei in both regions I and II (with relative amounts depending on the abundance of the paramagnetic spins). The

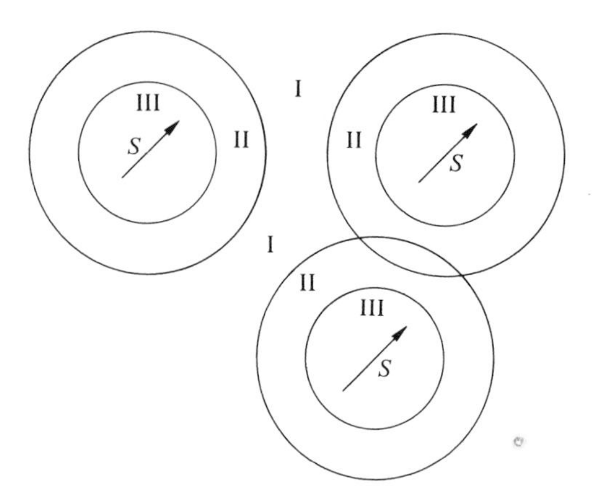

Fig. 4-16 Three separated paramagnetic impurity spins and the surrounding regions as defined in the text.

recovery of the signal is thus bound to be nonexponential in time, and the sample magnetization is clearly spatially inhomogeneous in the process. The nonexponential recovery of the magnetization is a clear indication of the spin-diffusion relaxation mechanism. Beyond this point, the problem is difficult to treat theoretically and is still the subject of some research.

If a nucleus has a quadrupole moment, it can relax by direct interaction with lattice vibrations (phonons). If the nucleus has a static quadrupole interaction, that interaction will be modulated by the distortions of the lattice caused by the vibrations. If the nucleus is at a site of cubic symmetry, and consequently has no static quadrupole interaction, distortions of the lattice by the vibrations remove the symmetry and produce fluctuating field gradients. There are two somewhat independent problems in understanding the relaxation mechanism. The first and most difficult is to understand the strength of the coupling to the lattice—to calculate the strength of the quadrupole interaction produced by the strain. The second is to understand the temperature dependence of the relaxation time. We shall confine ourselves here to a qualitative picture of the mechanism, without trying to make quantitative progress on either of the main problems.

Figure 4-17 shows the electric quadrupole interaction of a hypothetical nucleus as a function of displacement Δr from its equilibrium position. To be quite general, we have shown the interaction to have a value at $r = 0$, a slope, and curvature there. We see immediately from Fig. 4-17 that if the nucleus undergoes a sinusoidal oscillation at frequency ω, amplitude

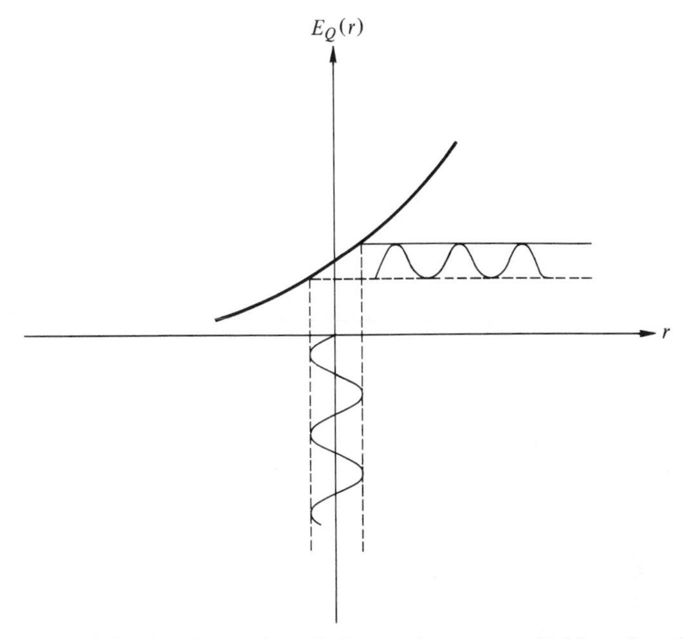

Fig. 4-17 Possible electric quadrupole interaction energy $E_Q(r)$ as function of displacement r from equilibrium position.

a_0, the quadrupole interaction energy is modulated at the same frequency by an amount

$$\delta E_Q = a_0 \left(\frac{\partial E_Q}{\partial r} \right)_{r=0} \cos \omega t \tag{4-40}$$

Equation (4-40) is a possible relaxation mechanism if $\omega = \omega_0$, a resonance frequency of the system. The efficacy of the mechanism depends on the probability that a lattice wave does indeed appear at frequency ω_0. If we assume a Debye model of the lattice, that probability depends on the energy density at $\nu_0 = \omega_0/2\pi$, and on the Bose–Einstein factor:

$$\rho(\nu_0)\, d\nu = \frac{12\pi}{c^3} \nu_0^2\, d\nu \frac{h\nu_0}{\exp(h\nu_0/kT) - 1} \tag{4-41}$$

where c is the velocity of sound. Since $h\nu_0/kT \ll 1$ for most practical cases, Eq. (4-41) reduces to

$$\rho(\nu_0)\, d\nu = \frac{12\pi}{c^3} kT\nu_0^2\, d\nu \tag{4-42}$$

We see immediately that the mechanism we have assumed (called the *direct* mechanism) is proportional to T. It is less obvious that it is very weak. The reason is the low density of modes at low frequency. In the more convenient quantum language, the mechanism we are discussing is the absorption or emission of a phonon at ν_0. The flux of phonons incident on the nucleus is proportional to ν_0^2. Since that flux continues to rise quadratically in the Debye model to the cutoff frequency ν_D, and since ν_D is typically 10^{12} to 10^{13} Hz, whereas ν_0 is seldom larger than 10^8 Hz for nuclei, most of the lattice modes are at much higher frequencies than ν_0, except at extremely low temperatures.

The next higher order process involves two lattice modes, rather than one, but both can be high frequency modes. The quantum mechanical description of this *Raman process* is phrased in terms of the inelastic scattering by the nucleus of a phonon of frequency ν to frequency ν', with the nuclear spin flipping in the process to absorb energy $h\nu_0 = h(\nu - \nu')$. The classical view of this process can be understood by referring again to Fig. 4-17. Let the two frequencies ν and ν' be simultaneously incident on the nucleus. Since the $E_Q(r)$ curve is not linear but has curvature, the time dependent quadrupole energy will have components not only at ν and ν', with amplitude proportional to $(\partial E_Q/\partial r)$, as before, but also at $\nu + \nu'$ and $\nu - \nu'$, with amplitude proportional to the curvature of the $E_Q(r)$ function at the origin: $(\partial^2 E_Q/\partial r^2)_{r=0}$. The probability of this two-phonon process depends on the simultaneous existence of two phonons; at high temperatures, then, one expects a T^2 temperature dependence. (High temperatures in the Raman case mean $T > T_D = h\nu_D/k$, so that all modes are excited with equal probability.) There seems to be no particularly simple way to show that at low T the Raman process goes as T^7. It becomes relatively ineffective, then, at low temperatures, where the advantage of the high density of modes at high frequencies is lost as their probability of excitation decreases rapidly with decreasing T.

Experimentally, it is hard to observe all the regions of temperature dependence. Paramagnetic impurities often dominate at low temperatures, and they also contribute a temperature dependent T_1 since T_{1S} is temperature dependent. In fact, the various low temperature regions, the direct, and the Raman, are most clearly observed in paramagnetic systems, where the detailed mechanisms are different, but the temperature dependence is still determined by the same properties of the lattice.

We note here, to end this section, that the first mechanism proposed and calculated for nuclear spin lattice was the modulation of the dipole-dipole interaction by the lattice vibrations. That such a mechanism would be relatively ineffective can be rendered plausible by the following. Compare Eq. (4-48), where the field gradient is determined by an isolated point charge of $+e$ at distance r_0, with the equivalent equation for the magnetic interaction between two dipoles μ separated by r_0. The ratio of the modulation of the magnetic to quadrupole coupling is

$$\frac{\delta E_H}{\delta E_Q} = \frac{(3\mu^2/r^4)\,\delta r}{(6e^2Q/r^4)\,\delta r} = \frac{\mu^2}{2e^2Q} \tag{4-43}$$

For typical values of $\mu = 10^{-24}$ ergs/G, and $Q = 10^{-24}$ cm^2, Eq. (4-43) comes out to be 2×10^{-6}. Since the relaxation rates will be in the ratio of the square of this ratio, the magnetic dipole mechanism is less effective by about 12 orders of magnitude!

4-6. SUMMARY AND LITERATURE SURVEY

The subject of resonance line shapes in a rigid lattice is rather complex. Our discussion here has been the merest outline. The most complete discussion for more advanced readers is to be found in Abragam [8], although Slichter [2] should be studied first for a detailed treatment of the algebra of the spin operators. Van Vleck's original paper [1] is rather hard going, but it does discuss comparison with experiment and theory. Later, better experiments [9] have been done, and the line shape, as well as the moments, have been calculated in CaF_2 (nuclear resonance of ^{19}F) and compared with experiment [9], [10]. But it must be admitted that there is a shortage of elementary material.

Elementary discussions are also in short supply on the subject of thermodynamics of spin systems. The review article by Hebel [11] is the most complete, and contains discussion of all but the more recent, exotic experiments.

The subject of quadrupole interactions in solids and molecules is likewise vast and quite devoid of elementary discussion. We have followed Slichter [2] rather closely, but have left out all the quantum mechanics. The first supplement to the *Solid State Physics Series* of Seitz and Turnbull by Das and Hahn [7] discusses pure quadrupole resonance. An article in the same series by Cohen and Reif [6] discusses quadrupole effects in nuclear magnetic resonance in the other limit, $\mathscr{H}_Z \gg \mathscr{H}_Q$. Neither article is elementary. Chapter 9 of Andrew [12] contains a summary of early work. A review of pure quadrupole resonance by its discoverer is to be found in the *American Journal of Physics* article by Dehmelt [13].

Problems

4-1. One of the important theorems that can be proven under rather general conditions is that the envelope of the free induction decay $F(t)$ following a 90° pulse is proportional to the Fourier transform of the absorption line shape $g(\nu)$. Defining $u = \nu - \nu_0$, the theorem for a symmetrical line is

$$F(t) = A \int_{-\infty}^{\infty} g(u) \cos ut \, du$$

where A is a constant.

(a) By expanding the right side of the preceding equation for $F(t)$ in a power series in t, show that the even moments of the line shape are given by the formula

$$\langle \Delta\nu^{2n} \rangle = (-1)^n \frac{(d^{2n}F(t)/dt^{2n})_{t=0}}{F(0)}$$

(b) Abragam [8] has given a convenient and remarkably accurate two-parameter formula for the free induction decay in a dipolar solid:

$$F(t) = \exp\left[-\left(\frac{a^2t^2}{2}\right)\right] \frac{\sin bt}{bt}$$

Show that the line shape $g(u)$ that corresponds to this form of $F(t)$ results from the superposition of Gaussian curves of root mean square half-width a, with a rectangular envelope of width $2b$.

(c) Show from the results of (a) and the expansion of $F(t)$ given in (b) in powers of t that the two parameters a and b are related to the second and fourth moments $\langle \Delta\nu^2 \rangle$ and $\langle \Delta\nu^4 \rangle$ by

$$\langle \Delta\nu^2 \rangle = a^2 + (\tfrac{1}{3})b^2$$

$$\langle \Delta\nu^4 \rangle = 3a^4 + 2a^2b^2 + (\tfrac{1}{5})b^4$$

4-2. To be a useful concept, spin temperature should usually behave in familiar ways. Thus, simple calorimetric experiments, such as allowing two spin systems at different temperatures to interact and equilibrate at a common spin temperature, should give the naïvely expected results. Consider the following sequence. A crystal of LiF is allowed to come to equilibrium in a field $H_0 \gg H_L$ at $T = T_L$. The ^{19}F system is then prepared by a 180° pulse, so that $T_F = -T_L$. The field is lowered to $H \gtrsim H_L$, where the cross-relaxation time is short, and the systems come to the same spin temperature, $T_F = T_{Li} = T_f$. Find T_f, and the magnitudes of the Li and F resonance signals relative to the respective signals before mixing. Assume the lithium is 100% ^{7}Li in the sample, rather than 94% ^{7}Li, 6% ^{6}Li, and that the spin-lattice relaxation times of all nuclei are very long compared with the duration of the experiment. (Many such experiments are described in Chapter 5 of Abragam [8].)

4-3. In Fig. 4-11, the most direct way to show that part (c) has the same interaction energy as (a) and (b) is to compute the electric field gradient at the origin. Show that all components V_{ij} of the electric field gradient vanish at the centers of a square and cubic arrays of identical point charges. (It is sufficient to show, in fact, that the contours of constant potential are circles and spheres near the origin for the square and cubic arrays, respectively.)

4-4. The orbital wavefunction of an electron in a p_z state may be written $\psi_{p_z} = f(r)z$, where $f(r)$ is a function only of the distance of the electron from the nucleus.

(a) Show that the component of the electric field gradient at the origin, $V_{zz}(0)$, produced by a unit charge at (x, y, z) is

$$V_{zz} = \frac{3z^2 - r^2}{r^5}$$

(b) Compute the total $V_{zz}(0)$ by weighing each point (x, y, z) by the charge density $e\,|\psi_{p_z}(x, y, z)|^2$ and integrating over all space. Show that the result is

$$V_{zz} = -|e|\,\frac{4}{15}\left\langle\frac{1}{r^3}\right\rangle$$

where

$$\left\langle\frac{1}{r^3}\right\rangle = \int_0^\infty \frac{1}{r^3}\,[rf(r)]^2 r^2\,dr$$

(c) The wavefunction for the $2p_z$ state of atomic hydrogen is $\psi = [1/(32\pi)^{1/2}]re^{-r/2}\cos\theta$, where r is expressed in units of the Bohr radius. Show that for deuterium, e^2qQ/h is about 50 kHz, where $Q = 2.77 \times 10^{-27}\ \text{cm}^2$. (For heavier atoms or ions, three factors combine to make e^2qQ/h in the tens of megahertz range, rather than tens of kilohertz: larger Q's, by as much as two orders of magnitude, larger $\langle 1/r^3\rangle$ because the nuclear charge is poorly screened for small r, and the Sternheimer antishielding factor.)

References

1. J. H. Van Vleck, *Phys. Rev.* **74**, 1168 (1948).
2. C. P. Slichter, *Principles of Magnetic Resonance*, Harper & Row, New York (1963).
3. N. Bloembergen, *Physica* **15**, 386 (1949). Reprinted in N. Bloembergen, *Nuclear Magnetic Relaxation*, W. A. Benjamin, Inc., New York (1961).
4. E. M. Purcell and R. V. Pound, *Phys. Rev.* **81**, 229 (1951).
5. A. G. Redfield, *Phys. Rev.* **98**, 1787 (1955). See also A. G. Redfield, *Phys. Rev.* **101**, 67 (1956) for the first application of adiabatic fast passage to solids.
6. M. H. Cohen and F. Reif, *Solid State Physics*, Academic Press, Inc., New York, vol. 5 (1957).
7. T. P. Das and E. L. Hahn, *Solid State Physics Supplement 1*, Academic Press, Inc., New York (1958).
8. A. Abragam, *The Principles of Nuclear Magnetism*, Oxford University Press, London (1961).
9. C. R. Bruce, *Phys. Rev.* **107**, 43 (1957).
10. I. J. Lowe and R. E. Norberg, *Phys. Rev.* **107**, 46 (1957).
11. L. C. Hebel, *Solid State Physics*, Academic Press, Inc., New York, vol. 14 (1963).
12. E. R. Andrew, *Nuclear Magnetic Resonance*, Cambridge University Press, Cambridge, England (1955).
13. H. G. Dehmelt, *Am. J. Phys.* **22**, 110 (1959), erratum, p. 317.

CHAPTER 5

Magnetic Resonance in the Alkali Metals

Some of the most important applications of magnetic resonance continue to be to the physics of metals. Because of their defining feature, a high density of conduction electrons, metals have certain special features as far as magnetic resonance is concerned, and as a result, resonance experiments have played an important role in understanding the properties of the electron gas in metals. The alkali metals are the simplest of all metals. Two of them, sodium and potassium, come close to satisfying the theorist's simplest model of a metal: an electron gas held together by a uniform background of positive charge. Almost all the presently conceivable magnetic resonance experiments have been done on lithium and sodium, and a relatively complete and internally consistent set of experimental data exists for these metals. In this chapter we shall use most of the qualitative and semiquantitative ideas of the first three chapters of this book in a description of these experiments and, hopefully, give some indication of the significance of the results.

5-1. THE PAULI SUSCEPTIBILITY OF AN ELECTRON GAS

At absolute zero, the electron gas in a metal is degenerate; all the electron states of the macroscopic "box" comprising the physical piece of metal are filled from the zero of energy until the number of available electrons is exhausted. Since the Pauli exclusion principle is obeyed, the electron states are filled with two electrons (of opposite spin) per state up to a maximum in energy, the Fermi energy. The Fermi energy is typically several electron volts, and the electron velocity is, from the relation $E_F = \frac{1}{2}mv_F^2$, about 10^8 cm/sec. At $T > 0°K$, the probability that a state of energy E is occupied is governed by the Fermi function

$$f(E) = \left[\exp \frac{(E - E_F)}{kT} + 1\right]^{-1} \tag{5-1}$$

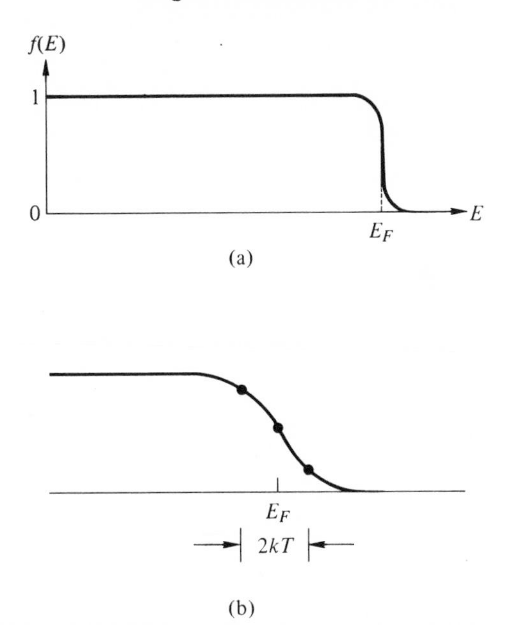

Fig. 5-1 (a) The Fermi function of some nonzero temperature. (b) The region near E_F on an expanded scale.

A graph of the Fermi function for some $T > 0°K$ is shown in Fig. 5-1. Note that it is unity until $E \cong E_F$, and falls from one to zero within an energy range of kT about E_F. One important consequence of this fact is that if a conduction electron is to participate in a process that changes its momentum or spin state with a change of energy of kT or less, it must be near the Fermi energy to begin in an occupied state and end in an unoccupied state. Thus, the only electrons that are free to participate in magnetic resonance are those at the very end of the Fermi distribution—a very small fraction, about kT/E_F, of the total. These electrons are also, then, nearly monoenergetic.

Let us apply a magnetic field to the sample. The energy of an electron may be written

$$E(H) = E(0) + |\gamma_e|\hbar m_S H_0 \tag{5-2}$$

where $E(0)$ is the kinetic energy in the absence of H_0. Note that we can divide the electrons into two groups, with different Fermi functions, one for $m_S = \frac{1}{2}$, the other for $m_S = -\frac{1}{2}$. The two distributions are shown in Fig. 5-2a, where it has been assumed that the field was turned on "instantaneously," and we are looking at the distribution before any spin flips have occurred. Thus half the spins have $m_S = \frac{1}{2}$ and half have $m_S = -\frac{1}{2}$. One can see immediately that the total energy of the system

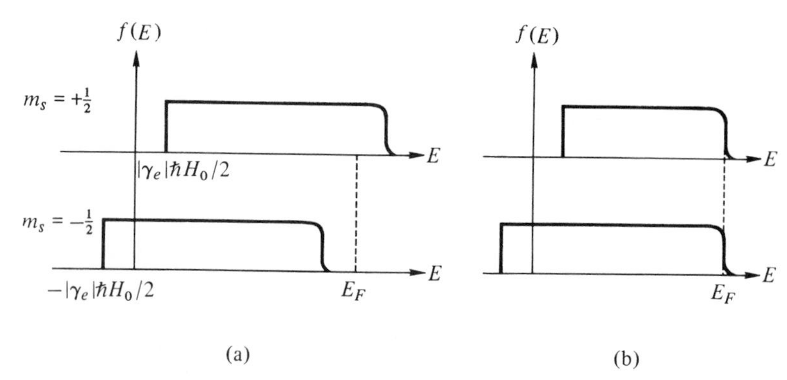

Fig. 5-2 (a) Fermi functions from $m_S = \pm\frac{1}{2}$ electrons immediately after H_0 is switched on, but before spin flips. (b) Equilibrium distribution: Fermi energies of the two spin states are equal.

is lowered if all the electrons above E_F in Fig. 5-2a, $m_S = -\frac{1}{2}$, are flipped to fill the empty states below E_F in the $m_S = +\frac{1}{2}$ distribution. The equilibrium, minimum energy situation is depicted in Fig. 5-2b, where we see that an unequal number of electrons are in the two spin states—a magnetic moment has developed. The total number of spins transferred from $-\frac{1}{2}$ to $+\frac{1}{2}$ was

$$\Delta n = \frac{N\gamma\hbar H_0}{2}\rho(E_F) \tag{5-3}$$

where $\rho(E_F)$ is the number of states per unit energy at the Fermi surface and N is the number of conduction electrons per unit volume. Normally, $\rho(E_F)$ is referred to as the density of states at the Fermi surface. (It is similar to the mode density of radiation in an enclosure, except electrons have mass, hence obey a different dispersion relation. The mode density goes as ν^2, the density of states as $\sqrt{E}$.) Each flipped spin changes the magnetic moment aligned along the field by twice its moment. Since it subtracts its moment from the $-\frac{1}{2}$ state and adds it to the $+\frac{1}{2}$ state, we get from Eq. (5-3) the total magnetic moment aligned along the field:

$$M = 2\,\Delta n\beta_0 = N\gamma\hbar\beta_0 H_0\rho(E_F) = 2N\beta_0{}^2 H_0\,\rho(E_F) \tag{5-4}$$

The Pauli susceptibility is thus

$$\chi_P = \frac{M}{H_0} = 2N\beta_0{}^2\rho(E_F) \tag{5-5}$$

Note that it is temperature independent, since we have made the tacit assumption that $kT \ll E_F$.

The alkali metals lithium and sodium are unusual among metals in that the conduction electron spin-resonance lines are quite narrow. That circumstance has made it possible to observe the electron resonance in these metals at low radio frequencies: 10^7 Hz instead of 10^{10}, or in fields of 3 G instead of 3000. Consequently, it has been possible to measure the Pauli susceptibility of lithium and sodium by a magnetic resonance technique. The idea is as follows. Recall that in Chapter 2, passing reference was made to the Kramers–Kronig relations, which relate to each other by integral expressions the susceptibilities $\chi'(\omega)$ and $\chi''(\omega)$. The quantity $\chi'(0)$ is just the static susceptibility, χ_n, the Curie law susceptibility for nuclei, or χ_P, the Pauli susceptibility for conduction electrons. The Kramers–Kronig relation at zero frequency states that $\chi'(0)$ is proportional to the area under the $\chi''(\omega)$ curve. The proportionality constant is the same whether it is the nuclear or electron spin resonance that is concerned. Hence, one may write the following ratio between χ_n and χ_P[1]:

$$\frac{\chi_n}{\chi_P} = \frac{\gamma_n}{\gamma_e}\frac{A_n}{A_e} \tag{5-6}$$

where A_n and A_e are the areas under the absorption curves of the nuclear and electron spin resonances. The ratio γ_n/γ_e appears because the experiment is done at constant frequency, and the field is swept to cover the resonances. Thus, one observes the nuclear resonance at 10^4 G with the field swept ± 10 G to cover that line, and the electron resonance at 3 G in the same apparatus, with the field swept from 0 to perhaps 20 G to cover that line. From Eq. (5-6), the Pauli susceptibility is thus measured in terms of the nuclear susceptibility χ_n, which is calculated from the completely reliable Curie law.

We shall quote the results of these measurements when they are needed to fit in with the many other resonance experiments.

5-2. THE ELECTRON-NUCLEAR INTERACTION

In most atoms, there is a hyperfine interaction between the valence electrons and the magnetic moment of the nucleus. The valence electron of alkali atoms is in an *s* state, so the hfs interaction depends only on $\psi(0)$, the probability amplitude for an electron to be at the nucleus at $r = 0$. It is convenient to write the interaction in terms of the nuclear and electron spin operators:

$$\mathscr{H} = a(s)\mathbf{I} \cdot \mathbf{S} \tag{5-7}$$

The coupling of Eq. (5-7) splits the ground state into the doublet of angular momentum $F = I \pm \frac{1}{2}$ with an energy separation of $a(s)(I + \frac{1}{2})$. This splitting, in the alkalis, ranges from 800 MHz in Li to about 9000 MHz in Cs.

The same interaction exists in the metal, but since the atomic valence electron is the one donated by each atom to the collection of itinerant conduction electrons, the s-electron wavefunction near the nucleus in the metal is not the same as in the atom; the difference is specified by a factor ξ, which is usually less than 1. Otherwise, the interaction is the same, so we can write, in the solid,

$$\mathscr{H} = a(s)\xi \mathbf{I}_i \cdot \mathbf{S}_j \tag{5-8}$$

as the interaction between the ith nucleus and the jth electron. The important question of the electron wavefunction is buried in the factor ξ, which, we shall see, may be measured. The interaction of Eq. (5-8) is responsible for the following list of effects on the electron and/or nuclear magnetic resonance: the Knight shift, nuclear relaxation and the Korringa relation, the Overhauser effect, and the electron resonance shift ("day shift"). We shall take up these effects in order.

The Knight shift

It will be our approach in this chapter to obtain answers the "quick and dirty" way. We begin by not being too stuffy about the subscripts i, j in Eq. (5-8). All nuclei are equivalent in simple monatomic metal, and we shall assume the Pauli principle keeps electrons out of the way of each other so that we can think of Eq. (5-8) alone as representing overwhelmingly the most important part of the nuclear interaction with all the conduction electrons. Put simply, a nucleus interacts with the electrons one at a time. Equation (5-8) may be written in the form

$$\mathscr{H} = \gamma_n \hbar I_z \, \Delta H \tag{5-9}$$

where ΔH is an internal effective field in the z direction caused by the S_z part of the hfs interaction. Comparison of Eqs. (5-8) and (5-9) gives, for ΔH,

$$\Delta H = \frac{\xi a(s)}{\gamma_n \hbar} \langle S_z \rangle \tag{5-10}$$

where $\langle S_z \rangle$ is the average value of the z component of the conduction electron angular momentum. We can identify $\langle S_z \rangle$ in terms of the static magnetization $M_0 = \chi_P H_0$, since we may also write $M_0 = N\langle S_z \rangle \gamma_e \hbar$, which yields

$$\langle S_z \rangle = \frac{\chi_P H_0}{N\gamma_e \hbar} = \frac{\chi_P \Omega H_0}{\gamma_e \hbar} \tag{5-11}$$

where N is the number density of electrons in the metal and $\Omega = 1/N$ is the atomic volume in the solid (assume one electron per atom). Equations (5-10) and (5-11) together give the Knight shift:

$$K \equiv \frac{\Delta H}{H_0} = \frac{\xi a(s) \chi_P \Omega}{\gamma_n \gamma_e \hbar^2} \tag{5-12}$$

It is instructive to digress for a moment and write the factor $\xi a(s)$ in the form

$$\xi a(s) = \frac{8\pi}{3} \gamma_e \gamma_n \hbar^2 \, |\psi_F(0)|^2 \tag{5-13}$$

Equation (5-13) was obtained from the free atom hfs interaction, which may be derived by a semiclassical argument (e.g., see Slichter [1], pp. 86ff.). In the free atom, the hfs for s states depends on $|\psi_A(0)^2|$, the squared amplitude in the atom of the s-state wavefunction at the nucleus. The subscript F on $|\psi_F(0)|^2$ in Eq. (5-13) is meant to indicate that in the metal $\psi(0)$ is to be evaluated for an electron on the Fermi surface. The quantity ξ is just given by

$$\xi = \frac{|\psi_F(0)|^2}{|\psi_A(0)|^2}$$

as implied by the comparison between Eqs. (5-7) and (5-8).

The Knight shift becomes

$$K = \frac{8\pi}{3} \chi_P [\Omega \, |\psi_F(0)|^2] \tag{5-14}$$

where we have singled out $\Omega|\psi_F(0)|^2$ in brackets to emphasize the following point. In a free electron theory of metals, in which the periodic potential produced by the ions is replaced by a uniform positively charged background, the electron wavefunctions are plane waves:

$$\psi_k = \Omega^{-1/2} \, e^{i\mathbf{k} \cdot \mathbf{r}} \tag{5-15}$$

The wavefunction in Eq. (5-15) is normalized to the volume occupied by an atom in the metal, Ω, which accounts for the $\Omega^{-1/2}$ in Eq. (5-15). The electron probability distribution is uniform: $|\psi_k(\mathbf{r})|^2 = |\psi_k(0)|^2 = \Omega^{-1}$. Hence, the factor in the brackets of Eq. (5-14) is unity. In a real metal, the ionic potential looks very much like the free atom potential near the nucleus, and *s*-like wavefunctions are strongly peaked at the nucleus. The electron probability density is thus strongly nonuniform, and the enhancement over the uniform conduction electron distribution given by Eq. (5-15) is just $[\Omega|\psi_F(0)|^2]$. The other two terms of Eq. (5-14) are now easily disposed of. The susceptibility, χ_P, is just the fractional polarization of the electrons, and the factor $8\pi/3$ comes from the demagnetizing factor of a uniformly magnetized sphere. The factor in brackets, $[\Omega|\psi_F(0)|^2]$, ranges from about 15 in lithium to about 2000 for a heavy metal such as mercury, where we find the 6*s* type wavefunctions to be strongly peaked at the nucleus once we get inside the screening charge of the core electrons and see the full charge of the 80 protons in the Hg nucleus.

The experimental situation is very much in accord with the picture presented here. In the simple metals (i.e., the alkalis and many others) the nuclear resonance *does* appear at a field higher than H_0 by KH_0, the amount predicted by Eq. (5-12). In fact, if Eq. (5-12) is the only consideration, the measurement of χ_P described previously, combined with the Knight shift, allows us to deduce the parameter ξ. There is excellent experimental evidence that the simple theory is, in fact, quite good, at least for the alkalis. The Knight shift ranges from very nearly zero in beryllium, 0.025% in lithium, all the way to 2.5% in mercury. The latter is a substantial shift indeed at 10 kG—about 250 G [2].

The relaxation time and the Korringa relation

The other parts of Eq. (5-8), the off diagonal terms $a(s)\xi(S_xI_x + S_yI_y)$, produce transverse fields that cause nuclear relaxation. They are, indeed, responsible for the nuclear spin-lattice relaxation in the alkali metals until, at least, the temperature comes close enough to the melting point to produce a motional nuclear dipole-dipole contribution to the relaxation, as in a liquid. If the correlation time of the conduction electrons is short enough (as seen by a nucleus), then the T_2^{-1} computed in Chapter 3 from random walk theory will also be equal to T_1^{-1}, since the interaction Eq. (5-8) is isotropic spatially.

The correlation time may be calculated from two equivalent points of view, both of them consistent with our insistence that on the average only one electron at a time occupies an atomic volume surrounding a nucleus. First, localize an electron to volume Ω at $t = 0$. Its wavefunction will spread substantially in a time $\hbar/E_F \simeq \tau_c$ [3]. Alternatively, we think of the electrons moving through the lattice at the Fermi velocity v_F, and ask how long it takes to travel a lattice distance, $\Omega^{1/3}$. The answer again is $\tau_c \simeq \hbar/E_F$. The precession rate of the nucleus about the hfs field is just $\delta\omega = \gamma_n \,\Delta H$, where ΔH is given by Eq. (5-9). Thus, our "random walk" estimate of the relaxation rate is, from Eq. (3-4),

$$T_1^{-1} = (\delta\omega)^2 \tau_c \alpha \tag{5-16}$$

The factor α is not in the original expression, Eq. (3-4). It is in Eq. (5-16) to take care of the fact that although there is on the average one conduction electron per nucleus in a monovalent metal, only those at the Fermi surface are available to interact. The electron's final state differs from the initial state by having a different wave vector as well as a different spin state, since the mutual spin flip described by Eq. (5-8) fails to conserve net Zeeman energy by the amount $(|\gamma_e - \gamma_n|)\hbar H_0$ and this energy must be taken up by the electron's kinetic energy. The whole interaction may be described as a spin-flip inelastic scattering of the electron by the nucleus.

We may estimate the fraction of electrons available to flip nuclear spins as

$$\alpha = \frac{kT}{E_F} \tag{5-17}$$

The factor $\delta\omega$ in Eq. (5-16) is the size of the instantaneous electron-nuclear interaction, or $\delta\omega = \xi a(s)/\hbar$. If we combine these quantities, we obtain

$$\frac{1}{T_1} = \left(\frac{\xi a(s)}{\hbar}\right)^2 \frac{\hbar k T}{E_F{}^2} \tag{5-18}$$

From Eq. (5-12) we can solve for $\xi a(s)$, to obtain the result for the relaxation time:

$$T_1 = \left(\frac{1}{K^2 T}\right) \frac{(\chi_P \Omega)^2 E_F{}^2}{k \gamma_e{}^2 \gamma_n{}^2 \hbar^3} \tag{5-19}$$

Equation (5-19) has two notable features. One is the inverse dependence of T_1 on T, which is normally expressed as $T_1 T = \text{constant}$. The other is the explicit expression of the fact that the hfs interaction is responsible for both the Knight shift and the nuclear relaxation. Equation (5-19), written as $K^2 T_1 T = \text{constant}$, is known as the Korringa relation.

Because our approximate derivation of Eq. (5-19) was done with a certain cunning when rough estimates of such factors as τ_c and α were made, Eq. (5-19) is only a factor $4\pi/9$ larger than the correct free electron expression. Equation (5-12), however, gives the correct free electron expression for the Knight shift. The proportionality of the relaxation rate to the absolute temperature is preserved in the more complete derivation, and has been checked for many metals, for some over the temperature range between their melting point and a fraction of a degree Kelvin.

It would not be consistent to check experimentally the Korringa relation in the form of Eq. (5-19) by using the experimental value of χ_P. Equation (5-19) was derived on the basis of a free electron model of a metal. The value of Knight shift, susceptibility, and relaxation time measurements is in the information they provide about how real metals differ from the simple model. A careful analysis of the experimental results for the alkali metals has recently been given by Narath [4]. His work shows that much remains to be learned beyond the free electron model, particularly with respect to the role of electron-electron interactions in the nuclear relaxation rate and the paramagnetic susceptibility.

The Overhauser effect

The various spin-lattice relaxation processes for conduction electrons were investigated theoretically by Overhauser [5] before conduction electron spin resonance had ever been observed experimentally. His investigations led him to propose [6] a method whereby nuclei may be made to polarize in magnetic field H_0 in a sample at temperature T as if they had the gyromagnetic ratio γ_e of the electron rather than γ_n of the nucleus. The Overhauser effect and many variations of it have been seen in many systems since it was originally proposed for metals; indeed, the nuclear Overhauser effect (i.e., between two nuclei of different γ's, rather than between nucleus and electron) has been useful in high resolution nuclear resonance spectroscopy.

We make the simplest explanation of the Overhauser effect by assuming electrons and nuclei of $I = \frac{1}{2}$ interact with the hfs interaction of Eq. (5-7). We assume further that the electrons obey Boltzmann statistics, and the high temperature approximation may be used. The energy-level diagram and the electron and nuclear m_z states are shown in Fig. 5-3, which illustrates the energy levels of the equation

$$E(m_S, m_I) = m_S \Delta - m_I \delta + A m_I m_S \tag{5-20}$$

where $\Delta = |\gamma_e|\hbar H_0$, $\delta = \gamma_n \hbar H_0$. In the figure it has been assumed that $\delta \gg A$, but that is not an essential assumption. The off diagonal terms of the hfs interaction produce the mutual spin-flip relaxation process

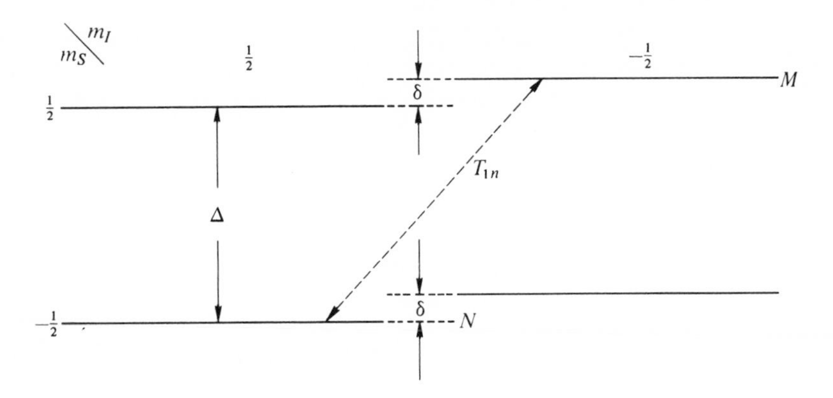

Fig. 5-3 Energy-level diagram of Eq. (5-20).

indicated by the dotted line and notation T_{1n} in Fig. 5-3. If we call the population of the $(m_S, m_n) = (-\frac{1}{2}, \frac{1}{2})$ state N, and the $(+\frac{1}{2}, -\frac{1}{2})$ state M. Then the thermal equilibrium population ratio, which is maintained by T_1, is, from Eq. (5-20),

$$\frac{N}{M} = \exp\left(\frac{\Delta + \delta}{kT}\right) \tag{5-21}$$

Now, if no other relaxation processes connect states of different nuclear spin [i.e., there is no process connecting $(\frac{1}{2}, \frac{1}{2})$ to $(-\frac{1}{2}, -\frac{1}{2})$], then we produce the Overhauser effect by applying enough rf power to saturate the purely electronic transitions. The situation is shown in Fig. 5-4. But nothing prevents the T_{1n} mechanism from continuing to maintain the same population ratio between the states it connects as it does in Eq. (5-21). We can then calculate the nuclear polarization, or, what is technically not

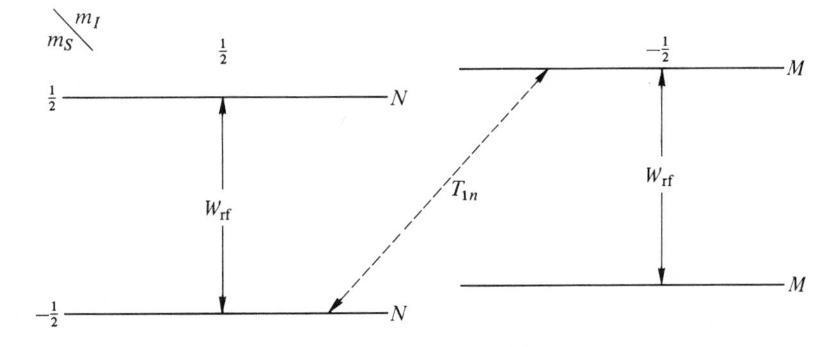

Fig. 5-4 Electronic transitions $\Delta m_S = \pm 1$ are saturated by W_{rf} at $\nu = \Delta/h$. The populations indicated produce nuclear polarization, the Overhauser effect.

quite the same thing, the ratio of $m_I = \frac{1}{2}$ to $m_I = -\frac{1}{2}$ populations. We need the total number of spins N_0, given by

$$2M + 2N = N_0 \tag{5-22}$$

The number of nuclei in the $m_I = \frac{1}{2}$ state is $2N$; the number of nuclei in the $m_I = -\frac{1}{2}$ state is $2M$. Thus, their ratio is

$$\frac{2N}{2M} = \exp\left(\frac{\Delta + \delta}{kT}\right) \simeq 1 + \frac{\Delta + \delta}{kT} \tag{5-23}$$

The number of nuclear spins up minus the number down is, from Eqs. (5-22) and (5-23),

$$2N - 2M = \tfrac{1}{2}N_0\left(\frac{\Delta + \delta}{kT}\right) \tag{5-24}$$

The thermal equilibrium population difference was $N_0\,\delta/2kT$, so the effect of saturating the resonance has been to increase the population ratio between the nuclear spin states by

$$\frac{\Delta + \delta}{\delta} = \frac{|\gamma_e|}{\gamma_n} + 1 \tag{5-25}$$

Overhauser's more general treatment gave the same result without the above restrictions; that is, Eq. (5-25) holds for any nuclear spin I, at arbitrary temperature, and for a degenerate conduction electron gas as well.

The most interesting application of the Overhauser effect (or rather, one related to it, the "solid state effect") has been to the polarization of nuclei to serve as targets for nuclear scattering experiments. This achievement took many years to accomplish, even though the Overhauser effect itself was first demonstrated not long after his original proposal.

As far as its usefulness to the study of metals is concerned, the Overhauser effect has had little practical application. One exception is that a careful study can reveal whether the full expected polarization has been obtained. If not, one concludes that other "short-circuit" nuclear relaxation processes exist in addition to the Korringa mechanism.

The electron resonance shift

The Knight shift argument, which leads to Eq. (5-9), may be turned around and applied to the electrons:

$$\mathscr{H} = \gamma_e \hbar S_z\, \Delta H' \tag{5-26}$$

where $\Delta H'$ is the internal field from the nuclei acting on the electrons. By analogy with Eq. (5-10),

$$\Delta H' = \frac{\xi a(s)}{\gamma_e \hbar} \langle I_z \rangle \tag{5-27}$$

where

$$\langle I_z \rangle = \frac{\chi_n \Omega H_0}{\gamma_n \hbar} \tag{5-28}$$

and χ_n is the usual Curie law nuclear susceptibility. If we write Eqs. (5-27) and (5-28) in terms of the Knight shift, we obtain for the fractional electron resonance shift

$$D = \frac{\Delta H'}{H_0} = K \frac{\chi_n}{\chi_P} \tag{5-29}$$

In fact, D is quite small. Although χ_n/χ_P depends on temperature, it would be difficult to increase it above 10^{-2}; therefore, for Na, for example, $D < 10^{-5}$, or a shift of 30 mG in 3000 G. It is very difficult to obtain CESR line widths as small as $\Delta H'$, so the measurement of D is quite difficult. It has been done for Na, however [7].

The nuclear susceptibility χ_n in Eq. (5-29) is proportional to—indeed, is a measure of—the nuclear polarization. If the electron resonance is saturated, χ_n must be multiplied by the Overhauser enhanced nuclear polarization, $s|\gamma_e|/\gamma_n$, where $0 < s < 1$ describes the degree of saturation of the electron resonance. Thus, D can be increased by about three orders of magnitude to 10^{-2}, a very substantial shift indeed. By using the Overhauser effect, D has been measured in Li [8]. The symbol D was chosen because of the somewhat waggish, but quite appropriate, observation that saturating the electron resonance makes the *Knight* shift disappear (since it destroys the electron polarization), and, via the Overhauser effect, causes a shift in the electron resonance, which we are not ashamed to refer to as the "day shift" [9].

It is not normally advisable to set down in a book of this nature the current experimental situation, complete with numbers. The facts have a way of changing, even when they seem firmly fixed. In the present case, temptation will not be resisted, even though the author is mindful of the dangers, because the experimental numbers *do* show so clearly the nature and extent of the contribution of magnetic resonance to the understanding of the behavior of electrons in simple metals. If we assume the theory to be correct, we can adopt the position that the experiments tell us about the two quantities χ_P and ξ. The former is of interest because the deviation of χ_P from the free electron value tells us something about electron-electron interactions in metals. The latter quantity, ξ, tells us something about the wavefunctions of electrons—it is connected more with the questions of the electron-lattice interaction, the band structure. The measured quantities are χ_P, K, D, and $a(s)$, which is known to great precision from atomic beam experiments. The spin-lattice relaxation time turns out to be less useful because the Korringa relation, Eq. (5-19), involves the square of so many experimental quantities—namely, K and χ_P—that even if the theory were above reproach, it would be a considerable experimental triumph to measure all these quantities to sufficient precision to make a really useful statement [4].

The three quantities, χ_P, K, and D, are measured by three independent experiments. We take the point of view that χ_P and K together determine ξ, as seen by Eq. (5-12). The electron resonance shift D also independently determines ξ. Equation (5-29) may be rewritten, with the aid of Eqs. (5-12) and (2-23), in the form

$$DT = \xi a(s) \frac{\gamma_n}{|\gamma_e|} \frac{I(I+1)}{3k} \tag{5-30}$$

Of course, $a(s)$ must be measured also, as must γ_e, γ_n, and Boltzmann's constant k. We may consider them all as known to infinite precision compared to the accuracy with which D can be measured. In Table (5-1), we have collected the experimental numbers for Li and Na, the only metals for which all the measurements have been done. The third column, $(\xi)_{K\chi}$, is inferred from the measured Knight shift and susceptibility from columns 1 and 2. The fifth column, $(\xi)_D$, is determined from Eq. (5-28). The sixth column gives the best theoretical values of ξ presently known. The comparison between $(\xi)_{K\chi}$ and $(\xi)_D$ shows the accuracy of the experiments and the self-consistency of the theory. Column six shows that the theory of the wavefunctions of electrons in Li and Na is in presentable shape.

5-3. CONDUCTION ELECTRON SPIN RESONANCE (CESR)

Magnetic resonance in metals is plagued by one complication not found in insulators: the rf skin depth. As a consequence of their high electrical conductivity, metals shield their interior from rapidly varying electromagnetic radiation, so that the rf fields needed to do a magnetic resonance experiment are confined to a layer on the surface of the metal, the so-called "skin depth." For radio frequencies in the 10-MHz range, the

Table 5-1

Resonance Shifts, Susceptibilities, and ξ for Li and Na

	K, %	χ_P, cgs volume units	$(\xi)_{K\chi}$	(DT), °K	$(\xi)_D$	$(\xi)_{\text{th}}$
Li	0.0249 ± 0.0005 [a,b]	$(2.08 \pm 0.1) \times 10^{-6}$ [d]	(0.43 ± 0.02)	$(0.99 \pm 0.03) \times 10^{-2}$ [f]	0.442 ± 0.015	0.49 ± 0.05 [g]
Na	0.1085 ± 0.0010 [a,c]	$(1.12 \pm 0.05) \times 10^{-6}$ [e]	(0.60 ± 0.03)	$(1.35 \pm 0.04) \times 10^{-5}$ [c]	0.63 ± 0.02	0.64 [h]

[a] [2]. [b] [10]. [c] [7]. [d] [1]. [e] [11]. [f] [8]. Value quoted is actually $DT[1 + s(|\gamma_e|/|\gamma_n|)]$ in the limit as $s \to 1$. [g] [12]. [h] [13].

skin depth δ of a good metal is on the order of 20×10^{-4} cm, or 20 μ (read "microns"). For microwave frequencies, say 10 GHz, δ is on the order of 1 μ at room temperature. In the previous discussion, we have avoided the complicating effects of the skin depth problem by assuming implicitly that all experiments are done on metal particles small in diameter compared with δ. Most nuclear resonance experiments on metals have, in fact, been carried out under these circumstances, and some conduction electron spin resonance (CESR) experiments at low frequencies have also. However, CESR in the low megahertz frequency range can be observed only if the relaxation times $T_1(=T_2)$ are so long that the line width is no greater than the field for resonance. It appears that only lithium at room temperature and sodium at liquid nitrogen temperature (77°K) and below, and possibly very pure potassium, satisfy the requirements that $T_1 \lesssim 10^{-8}$ sec, which implies a line width on the order of 10 G. It is more conventional to do electron resonance in fields of a few thousand gauss, say 3000 G, with the Larmor frequency around 9 GHz. Under these conditions, although the rf skin depth is on the order of 1 μ at room temperature for a good conductor, it becomes even less as the conductivity increases at lower temperatures. The theoretical simplification achieved by doing the resonance on particles small compared with the skin depth requires samples too small to prepare easily, and it becomes necessary to examine the consequences of observing the resonance in samples large compared with the skin depth. The complications introduced by this consideration turn out to have new physics in them, and one gains much more by doing experiments on large samples than was obvious before all the ramifications were discovered.

It must be mentioned by way of beginning that it is possible to do a nuclear resonance in large metallic samples; indeed, one can observe a perfectly acceptable signal in a macroscopic cylinder of metal with a coil of wire wrapped around it. The difference between the metallic and the dielectric sample is that the rf field dies away exponentially in the metallic sample with the $1/e$ attenuation length of $\delta = c/(2\pi\omega\mu\sigma)^{1/2}$, where c is the velocity of light, ω the frequency, μ the relative permeability, and σ the conductivity (in esu). The only signal comes from the nuclei within roughly δ of the surface, and its magnitude may be computed by determining the surface impedance $\mathscr{Z}_s = \{(4\pi/c)[\mathbf{E}(0) \times \mathbf{H}(0)]/H(0)^2\} \cdot \mathbf{n}$, where $\mathbf{n}$ is the unit vector normal to the surface of the sample. The argument "0" indicates the value of the fields at the surface of the sample. The calculation is simply a boundary value problem involving Maxwell's equations and the correct boundary conditions on $\mathbf{E}$ and $\mathbf{H}$ at the surface of the conductor. The resonance aspect of the problem is introduced by writing the permeability $\mu = 1 + 4\pi(\chi' - i\chi'')$. For a spectrometer tuned to be sensitive to absorbed power, the "lossy" or real part of the complex surface impedance $\mathscr{Z}_s$ includes a term proportional to $(\chi' + \chi'')$. Thus, the line shape in a "thick" metallic sample turns out to be this particular mixture of the real and imaginary parts of the complex susceptibility—it is an asymmetric line shape.

The difference between conduction electrons and the preceding example is that the former are mobile whereas the latter were tacitly assumed to be static. That the difference is a fundamental one can be seen by making a simple random walk estimate of the rms distance a conduction electron diffuses during a transverse relaxation time T_2. The step length in the problem is given by

$$\Lambda = v_F \tau \tag{5-31}$$

where v_F is the Fermi velocity and τ is the mean time between collisions of the electron with the "lattice." That time is the "relaxation time," which determines electrical conductivity,

$$\sigma = \frac{Ne^2\tau}{m} \tag{5-32}$$

where N is the number density of conduction electrons of charge e, mass m. τ is governed by electron collisions with impurities or sample boundaries at very low temperatures in very high purity metals, and by electron-phonon scattering at high temperatures (such as room temperature) in metals even of moderate purity. The range of τ is from roughly as long as 10^{-8} sec at low temperatures to 10^{-14} sec for a fair conductor at room temperature. The Fermi velocity appears in Eq. (5-31) because the electrons whose spins are flipped in the CESR experiment are at the top of the Fermi distribution. v_F is typically about 10^8 cm/sec.

The random walk estimate of the diffusion distance in T_2 is thus

$$\langle d^2 \rangle \simeq \Lambda^2 \frac{T_2}{\tau} = v_F{}^2 \tau T_2 \simeq \mathscr{D} T_2 \tag{5-33}$$

or, a rms distance of

$$d_{\text{rms}} \simeq (\mathscr{D} T_2)^{1/2} \tag{5-34}$$

The diffusion constant $\mathscr{D}$, introduced in Eq. (5-33), is roughly given by

$$\mathscr{D} \sim v_F{}^2\, \tau = v_F \Lambda \tag{5-35}$$

where $\mathscr{D}$ is typically on the order of 100 cm^2/sec for conduction electrons at room temperature ($\Lambda \sim 100$ Å, $v_F \sim 10^8$ cm/sec). If T_2 is 10^{-7} sec, a value easily achieved in moderately pure lithium at room temperature, then d_{rms} is about 30 μ, roughly an order of magnitude larger than the skin depth. (Factors on the order of unity have been disregarded in both the calculations and the definition of $\mathscr{D}$. Only the order of magnitude of the result is required.)

Thus the picture we must have for the rapidly diffusing conduction electrons is one of spins randomly diffusing in and out of the skin depth many times during a phase coherence time. The time it takes for the electron to diffuse through the skin depth is

$$t \simeq \frac{\delta^2}{2\mathscr{D}} \tag{5-36}$$

which is on the order of 10^{-10} sec for Li at room temperature. That time is a microwave period at 3 cm wavelength. A quick application of the uncertainty principle, or, what is the same thing, a quick calculation of the Fourier transform of a one-period pulse of radio frequency at ω, gives a spectral width of ω. Therefore, the line should be as broad as $1/t$ of Eq. (5-36).

The conclusion is incorrect. There will be many electrons which, in their random walk, will come back into the skin depth many times in a phase coherence time T_2. They will have preserved their phase memory if they have not relaxed, and the phase of the spin precession relative to the phase of the rf field will be constant. The situation is quite analogous to the Ramsey split-field resonance experiment discussed in Chapter 1, except that the time spent *out* of the rf field will be randomly distributed among the spins. All that will remain will be the sharp central peak of width $\Delta\omega \sim 1/T_2$. In terms of Fourier transform language, the rf field seen by a typical spin is not a single short pulse, but a series of pulses random in time but coherent in phase, with the series lasting a time T_2 for an average spin. The Fourier transform of that situation is a sharp ($\Delta\omega \sim 1/T_2$) peak at ω and a flat, weak background of width $1/t$. Figure 5-5 illustrates these remarks schematically. The broad background absorption is, of course, not seen, but the central sharp resonance is, and its width is the same as if no diffusion were occurring. However, the shape of the resonance does depend on the existence of diffusion, and is, in fact, different from its form in the static spin case. The shape also depends on the sample size in intermediate cases where the thickness of the sample is not large compared to the diffusion length d_{rms} of Eq. (5-34).

The theory of the CESR line shapes for a flat metallic plate of arbitrary thickness was first worked out by Dyson [14], and the particular asymmetry of the lines is often referred to as the "Dyson shape." Dyson's theory was recast in more familiar language by Kaplan [15], who re-

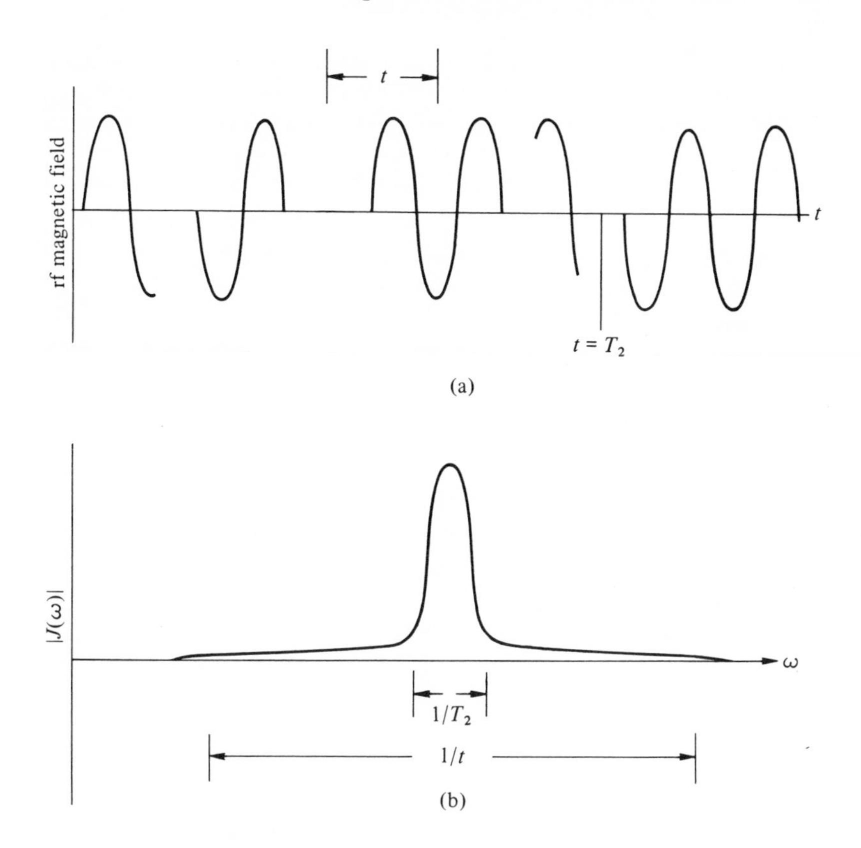

Fig. 5-5 (a) Radiofrequency fields as seen by an electron diffusing in and out of the skin depth in a metal. (b) Spectral density $J(\omega)$ of the rf field seen by the diffusing electron.

derived Dyson's results using the Bloch equations as a starting point. It had long been known that in a liquid, such as water, the decay of the echo envelope in Hahn's spin-echo experiment (Chapter 2) was often not exponential, as predicted by the Bloch equations. The source of the difficulty is in the ability of the nuclei in the liquid to diffuse in the external inhomogeneous magnetic field to a rather different value of the field during the time, $t \sim T_2$, between rf pulses. The consequent change of an individual nucleus's Larmor frequency during the course of the experiment is not taken into account in the simple Bloch equation description of the spin-echo decay. The effect can be included in the theory, as shown by Torrey [16], by adding a term to the Bloch equations of form $\mathscr{D}\nabla^2 M$, where $\mathscr{D}$ is the diffusion constant, and ∇^2 is the Laplacian operator. The addition of $\mathscr{D}\nabla^2 M$ is plausible since it, by itself, is the spatial contribution to dM/dt.[1] Kaplan [15] simply showed that the Bloch equations with diffusion, plus Maxwell's equations, and the proper boundary conditions at the surface of the metal, could be solved to yield Dyson's line shape. Of course, the spatially inhomogeneous transverse magnetization that exists if $\nabla^2 M$ is not to be zero is not caused by an external static magnetic field inhomogeneity in this case. There is a gradient in the transverse magnetization because the external rf field dies away in a skin depth, and the transverse magnetization is thus also spatially inhomogeneous.

The transverse magnetization does not die away in an rf skin depth, however, but persists into the interior of the metal a distance on the order of the diffusion length d_{rms}. The consequence of this fact was pointed out by the Russian physicist Azbel' [17], who showed that in a flat metal plate of thickness comparable to $d_{\text{rms}} \gg \delta$, electromagnetic energy would be transmitted from one side to the other if the CESR condition was satisfied. What happens is that the macroscopic magnetization, created by the spins that maintain their coherent admixture of spin up and spin down states for a distance d_{rms} inside the metal, induces a microwave current on the far side of the sample, and the microwave power radiated by that current can be detected by a sensitive microwave receiver. The geometry of a CESR transmission experiment is indicated in Fig. 5-6.

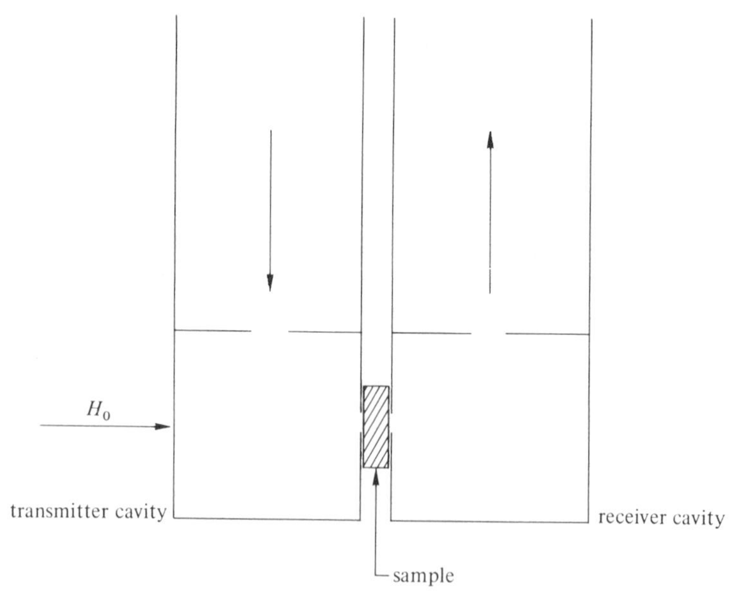

Fig. 5-6 Schematic diagram of a CESR transmission experiment.

[1] Remember that the diffusion equation governing the time rate of change of a spatially inhomogeneous quantity, in this case M, is $\partial M/\partial t = \mathscr{D}\nabla^2 M$.

In addition to being a very pretty experiment [18], [19], the transmission experiment has made possible the observation of conduction electron resonance in some metals other than the alkalis [20]. The problem had always been that it was difficult to tell whether a resonance in a metal came from the conduction electrons or from paramagnetic impurities fixed within the sample. The transmission experiment discriminates strongly against fixed spins, so the source of the resonance can be clearly determined.

A more important application came to light when it was found [21] that at very low temperatures a transmission signal in sodium and potassium occurred not only at the field satisfying the usual resonance condition, but also at several discrete fields displaced a few gauss from the main resonance field. The condition satisfied by the sample in order to produce these effects was that the resistivity relaxation time τ was so long that $\omega_c \tau > 1$, where ω_c is the cyclotron frequency of the electrons. That is, the sample had to be so pure and the temperature so low that an electron would complete one or more circular orbits of frequency

$$\omega_c = \frac{|e| H_0}{mc} \tag{5-37}$$

between collisions. Under those conditions, it turns out that the diffusion constant $\mathscr{D}$ has an imaginary part—it is not a real number. The "side-band resonances" were given the name "spin waves."[2] The imaginary part of $\mathscr{D}$ has been shown to result from the spin-dependent part of the electron-electron interactions in the metal, just the same interaction that causes the Pauli susceptibility to be somewhat enhanced over the free electron value of Eq. (5-5). We have the very gratifying result that CESR provides two independent ways of measuring a property of the degenerate electron gas which is of fundamental interest. At this writing, the enhancements of the susceptibility measured by the two methods are in fair agreement with each other.

Other basic properties of CESR of particular interest are the g shift and the spin-lattice relaxation time. Unfortunately, the theory of both of these properties is quite complicated and difficult to summarize simply. Both depend on the spin-orbit interaction, but since the conduction electron is not localized, as it is in paramagnetic salts, onto a single

[2] A name perhaps chosen a little for its shock value, since spin waves are the elementary excitations of a ferromagnet and antiferromagnet, in the same sense that phonons are the elementary excitations of the elastic lattice. The name has its justification, however, as one can see if he remembers that Schrödinger's equation for a free particle, which certainly has wavelike solutions, is the diffusion equation with a pure imaginary diffusion constant. The alkali metals are, of course, not ferromagnetic.

centro-symmetric potential, the theoretical description becomes unwieldy. An advanced, no-holds-barred review of the theoretical situation up to 1963 may be found in the article by Yafet [22].

5-4. LITERATURE SURVEY

Most elementary, or introductory essays on nuclear magnetic resonance contain a discussion of the special phenomena characteristic of metals, particularly the Knight shift. The advanced literature, even the review literature, is vast and quite specialized, particularly when problems of alloys are discussed. The earliest comprehensive review of the data and the theory, still the most frequently cited reference in the field, is the review article by Knight [2]. The best known review of alloy data—a subject not at all dealt with in this chapter—is that of Rowland [23]. It is obviously not current. The literature of conduction electron spin resonance in metals is very small, reflecting the relative inactivity of the field for many years after its initial discovery. The recent burst of activity following the introduction of the transmission technique is unfortunately at this writing still almost entirely in the form of letters to the *Physical Review*, not a medium often used for elementary expositions of any subject. It is perhaps too early to say whether the field will develop enough to merit extensive coverage in the review literature.

Problems

5-1. Use the principle of detailed balance to derive the Overhauser effect for the system of Fig. 5-4 with incomplete saturation of the electronic transitions. Show that the population ratio of the nuclear spin states is increased by

$$\left[\frac{s\,|\gamma_e|}{\gamma_n}+1\right]$$

instead of Eq. (5-25). The factor s is given by $s = [S_0 - \langle S_z\rangle]/S_0$, where S_0 is the equilibrium electronic polarization and $\langle S_z\rangle$ the electronic polarization in the presence of W_{rf}. The derivation in the text assumed $\langle S_z\rangle = 0$, or $s = 1$.

References

1. C. P. Slichter, *Principles of Magnetic Resonance*, Harper & Row, New York (1963).
2. A survey of the basic work on the Knight shift is given by W. D. Knight, *Solid State Physics*, F. Seitz and D. Turnbull, Eds., Academic Press, Inc. New York (1955), pp. 93ff.

3. D. S. Saxon, *Elementary Quantum Mechanics*, Holden-Day, Inc., San Francisco (1968), pp. 64ff.
4. A. Narath and H. T. Weaver, *Phys. Rev.* **175**, 383 (1968).
5. A. W. Overhauser, *Phys. Rev.* **89**, 689 (1951).
6. A. W. Overhauser, *Phys. Rev.* **92**, 411 (1953).
7. C. Ryter, *Phys. Letters* **4**, 69 (1963).
8. C. Ryter, *Phys. Rev. Letters* **5**, 10 (1960).
9. G. Pake, *Solid State Physics*, F. Seitz and D. Turnbull, Eds., Academic Press, Inc., New York (1955), p. 91.
10. R. T. Schumacher and N. S. VanderVen, *Phys. Rev.* **144**, 327 (1966).
11. R. T. Schumacher and W. E. Vehse, *J. Phys. Chem. Solids* **24**, 297 (1963).
12. T. Kjeldaas and W. Kohn, *Phys. Rev.* **101**, 66 (1956).
13. R. Taylor, R. A. Moore, and S. H. Vosko, *Can. J. Phys.* **44**, 1995 (1966).
14. F. J. Dyson, *Phys. Rev.* **98**, 349 (1955).
15. J. Kaplan, *Phys. Rev.* **115**, 575 (1959).
16. H. C. Torrey, *Phys. Rev.* **104**, 563 (1956).
17. M. Ya. Azbel' and I. M. Lifshitz, in *Progress in Low Temperature Physics*, C. J. Gorter, Ed., North Holland Publishing Company, Amsterdam, vol. 3 (1961), p. 329.
18. R. B. Lewis and T. R. Carver, *Phys. Rev. Letters* **12**, 693 (1964).
19. N. S. VanderVen and R. T. Schumacher, *Phys. Rev. Letters* **12**, 695 (1964).
20. S. Schultz and C. Latham, *Phys. Rev. Letters* **15**, 148 (1965); S. Schultz and M. R. Shanabarger, *Phys. Rev. Letters* **16**, 178 (1966).
21. S. Schultz and G. Dunifer, *Phys. Rev. Letters* **18**, 283 (1967).
22. Y. Yafet, in *Solid State Physics*, F. Seitz and D. Turnbull, Eds., Academic Press Inc., New York, vol. 14 (1963).
23. T. J. Rowland, *Progress in Materials Science*, Pergamon Press, London, vol. 9 (1961), pp. 1–93.

CHAPTER 6

Miscellaneous Subjects

The resonance technique has great versatility. It finds application in a wide variety of contexts that cannot simply be described as the production and detection of a macroscopic transverse magnetization. In this chapter, we shall discuss some experiments that have either had considerable application in the past or seem to have a very promising future, but that do not fit into the pattern of the previous chapters. From among the many subjects that might reasonably find their way into this discussion I have chosen cyclotron resonance, optical pumping, and some experiments on the resonance of excited states, both atomic and nuclear. All the techniques have applications in solid state physics, atomic physics, nuclear physics, and chemistry.

6-1. CYCLOTRON RESONANCE

For convenience, we repeat the introductory development of Chapter 1. Consider a charged particle moving with velocity **v** in a uniform **B**. The Lorentz force on the body is the familiar expression

$$\mathbf{F} = \frac{q}{c}\mathbf{v} \times \mathbf{B} \tag{6-1}$$

where **F** produces the acceleration

$$\frac{d^2\mathbf{r}}{dt^2} = \frac{q}{mc}\mathbf{v} \times \mathbf{B} \tag{6-2}$$

of the mass m. Since the acceleration is perpendicular to **v** (and **B**), the motion has a constant velocity component $v_{\|}$ along **B** and a constant acceleration component

$$a_{\perp} = \frac{q}{mc}v_{\perp}B = \frac{{v_{\perp}}^2}{r} \tag{6-3}$$

perpendicular to **B**. The resultant motion projected on the plane perpendicular to **B** is a circle of radius r:

$$r = \frac{mc}{qB} v_{\perp} \tag{6-4}$$

The particle has an angular orbital rotation frequency

$$\omega_c = \frac{v_{\perp}}{r} = \frac{qB}{mc} = \gamma_c B \tag{6-5}$$

where ω_c is called the cyclotron frequency.

The original cyclotron of Lawrence can be regarded as a device for increasing the cyclotron radius r, and hence the kinetic energy, by synchronous applications of an electric field tangential to the orbit. Figure 6-1a shows that familiar device with the "dees" that provided the tangential electric field. Figure 6-1b shows an easier way of doing the same thing with the electric field applied to the capacitor plates. Although this method slightly distorts the circular orbit, it clearly has the same qualitative effect. Figure 6-1c shows one of the rotation components of **E**, which provides a constant tangential accelerating force on the particle. The counterrotating component produces no effect other than a slight distortion of the orbit. The same discussion was made in the case of the linearly polarized rf magnetic field (Chapter 1).

In the applications of interest to us, the initial velocities are often due to thermal motion. Our object will not be to accelerate the particles to high energy but to use the device as a mass spectrometer; that is, the experiments measure ω_c in a known field B, and hence determine m in Eq. (6-5). So the final velocities after acceleration are generally not much larger than thermal, and the final radii are still small. For this, the cyclotron arrangement of Fig. 6-1a is impractical, and the experiments with which we are concerned use experimental arrangements such as Fig. 6-1b.

We want to develop equations of motion for an assembly of n charged particles per cubic centimeter, charge q, in a magnetic field B_0, under the influence of an accelerating field

$$\mathbf{E} = E_0[\hat{\mathbf{i}} \cos \omega t - \hat{\mathbf{j}} \sin \omega t] \tag{6-6}$$

(The sense of rotation of **E** has been chosen to make it synchronous for $q > 0$.) We also wish to introduce into the problem a scattering or relaxation mechanism that scatters the particles out of their orbits at a

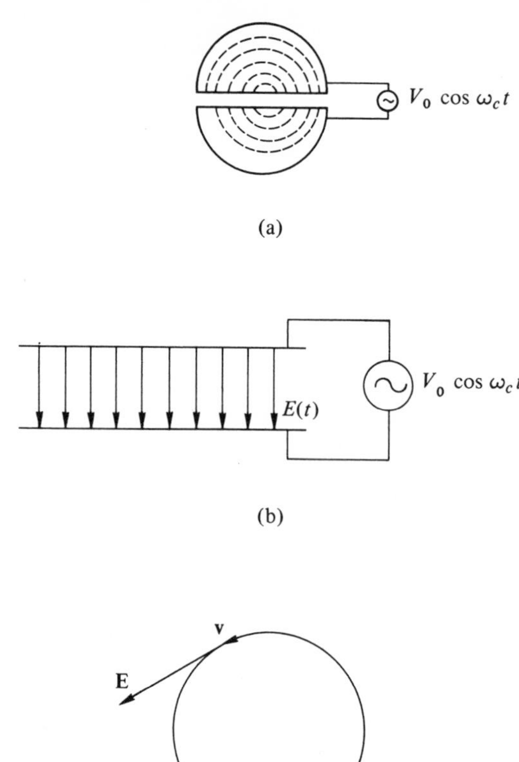

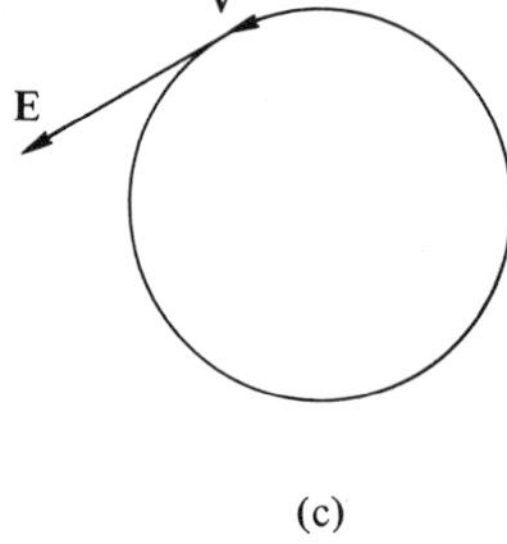

Fig. 6-1 (a) The "dees" of a cyclotron and the spiral path of an accelerated particle. The magnetic field is perpendicular to the page. (b) Parallel plate capacitor method of cyclotron acceleration. (c) Rotating frame view of accelerating field produced in (b).

rate $\nu_r = 1/\tau_r$. The differential equation for the average velocity $\mathbf{v}$ of an ensemble of particles is

$$\frac{d\mathbf{v}}{dt} = \frac{1}{m} q\mathbf{E}(t) + \frac{q}{mc} \mathbf{v} \times \mathbf{B}_0 - \mathbf{v}\nu_r \tag{6-7}$$

The relaxation term has been given the form in Eq. (6-7) by analogy with resistivity and other viscous damping phenomena. It is convenient to work with the macroscopic current density

$$\mathbf{J} = nq\mathbf{v} \tag{6-8}$$

where n is the number density of particles with velocity $\mathbf{v}$. Equation (6-7) may then be recast into

$$\frac{d\mathbf{J}}{dt} = \frac{nq^2}{m}\mathbf{E}(t) + \frac{q}{mc}\mathbf{J} \times \mathbf{B}_0 - \frac{\mathbf{J}}{\tau_r} \tag{6-9}$$

Equation (6-9) bears a striking resemblance to the Bloch equations for the magnetic moment, except for the form of the driving term $nq^2E(t)/m$. In the steady state we may solve for the power absorbed from the driving field by the same steps that were followed for the Bloch equations to obtain

$$P(\omega) = \mathbf{J} \cdot \mathbf{E}(t) = \frac{nq^2\tau}{m} \frac{E_0{}^2}{1 + (\omega - \omega_c)^2} \tau_r{}^2 \tag{6-10}$$

Equation (6-10) is a Lorentz curve centered on ω_c with a half-width at half-maximum of τ_r^{-1}.

The simplest application of these ideas is in a device known originally as an "omegatron" [1], and, more recently, as the ion cyclotron [2]. In its original form, it was used to measure the cyclotron frequency of the proton in the same field B_0 in which the nuclear magnetic resonance frequency of protons in H_2O was measured. The ratio of the two frequency measurements gives the intrinsic magnetic moment of the proton in terms of the Bohr magneton. The proton moment is an important physical constant and plays an important role in fixing the values of the fundamental constants.

The ion cyclotron is a recent development that has so far been applied principally to chemical problems. A block diagram of the ion-cyclotron apparatus is shown in Fig. 6-2. Notice that the components used are all standard in magnetic resonance. The spectrometer may be a marginal oscillator or a Robinson circuit (see Chapter 2), and the auxiliary detecting and recording equipment are the same as for any nuclear magnetic resonance experiment. The difference is that the power is absorbed from the electric field in the capacitor of the tank circuit by the electric dipole moment of the circulating charged particle, rather than from the magnetic field of the inductor by the magnetic dipoles.

Some additional insight into the operation of the ion cyclotron is obtained by considering the effect of the alternating electric field on two particles (Fig. 6-3) that are out of phase with each other. Particle 1 has the proper phase to be accelerated by the applied rf electric field in both (a) and (b), which is for a half-period later. Particle 2 is decelerated. Hence, the magnitude of the rotating electric dipole moment p_1 is increased,

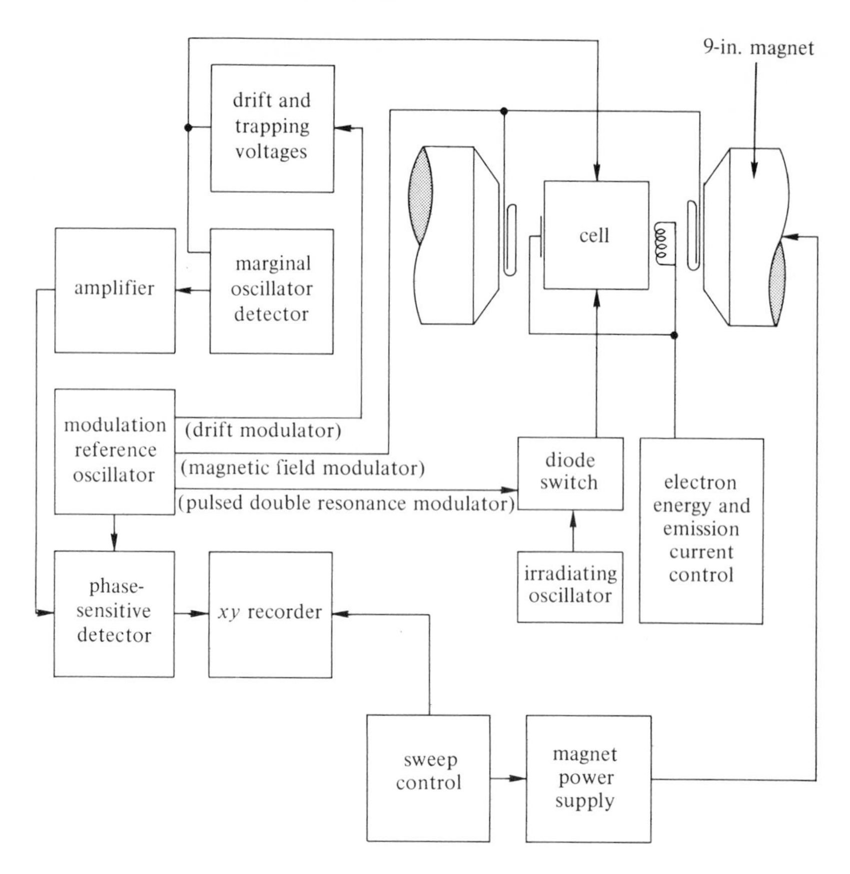

Fig. 6-2 Bloch diagram of ion-cyclotron resonance spectrometer (after Baldeschwieler [2]).

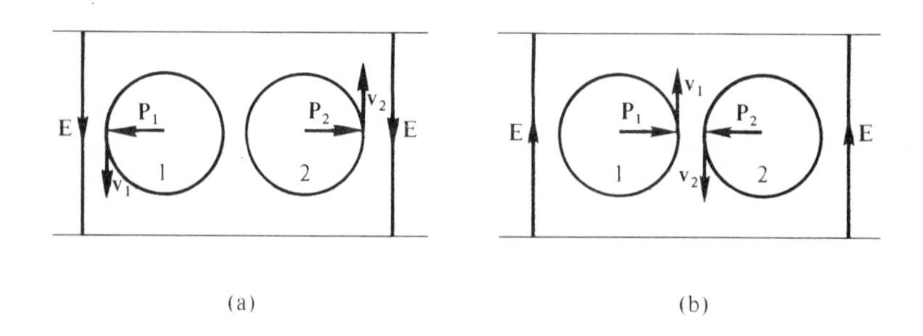

Fig. 6-3 The two charged particles, 1 and 2, are 180° out of phase with each other in their cyclotron orbits. Particle 1 is accelerated by the rf field both in (a), at time t, and in (b), at $t + T_c/2$. Particle 2 is decelerated at both times.

and p_2 is decreased. Thus the action of the rf field is to decrease the energy and the electric dipole moment of out-of-phase particles. Once we see how the phase coherence is imposed on the system in this way, it is not difficult to use our understanding of the magnetic moment problem to comprehend the behavior of the ion cyclotron. As an example, spin-echo effects may be expected, as suggested by the fact that Eq. (6-9) for **J** is, without the driving term, exactly the same as the Bloch equation for **M**.

We can also estimate the signal power from Eq. (6-10), at $\omega = \omega_c$:

$$\frac{P}{n} = \frac{q^2 \tau_r}{m} \left(\frac{V}{d}\right)^2 \tag{6-11}$$

where V is the voltage applied to the plates of the capacitor, and d is the plate separation. If we guess $\tau_r = 10^{-4}$ sec, $d = 1$ cm, $V = 1$ V (1/300 statvolt), and $m = 50$ proton masses, we find P/n to be 0.3×10^{-12} W. If the capacitor is part of a high Q tank circuit, that power might easily be developed across $10^4\ \Omega$ in a practical circuit. If so, that power per ion per cubic centimeter corresponds to a signal voltage of about 50 μV, an easily detectable signal. The estimate is suspect because it suggests that the resonance from one ion is easily detectable, although Eq. (6-10) is derived from macroscopic equations that are not valid for densities down to one ion per cubic centimeter. Nevertheless, the calculation suggests that sensitivities to that level can be expected.

The ion-cyclotron resonance experiment can take advantage of both the mass spectrometry capabilities of the instrument and the ability to measure τ_r by measuring the line width of the resonance. The former use is obvious: If several ions of different mass are present in the capacitor region, one may selectively detect the ion-cyclotron resonance of an ion of mass m and not of mass m' by applying the frequency $\omega_c = eB/mc$ and not $\omega_c' = eB/m'c$. To see why τ_r might be interesting, we must discuss possible relaxation mechanisms. The experiment must, of course, be done in high vacuum so that the scattering of the ions of interest from residual gas molecules is negligible. Thus, we require $\omega_c \tau_r \gg 1$ or an ion mean path $\Lambda = \bar{v}\tau_r \gg 2\pi\bar{v}/\omega_c$ where $\bar{v}$ is the ion velocity. The relation between the mean free path Λ, the foreign or residual gas number density N, and the collision cross section σ is

$$\Lambda^{-1} = N\sigma \tag{6-12}$$

The number density N must then satisfy the inequality

$$N \ll \frac{\omega_c}{2\pi\bar{v}\sigma} \tag{6-13}$$

On the other hand, one may use the observed width of the resonance to measure the collision cross section σ between the ion whose resonance is being observed and another molecule or ion. The cross section σ is given by the expression

$$\sigma = \frac{1}{N v_{\text{rel}} \tau_r} \tag{6-14}$$

where $1/\tau_r$ is, as before, the half-width of the resonance, and N and v_{rel} must be known. v_{rel} is the relative velocity of the colliding particles.

More complex and more interesting phenomena can be studied if one pursues more carefully the arguments leading to the Bloch-like equation for **J**, Eq. (6-9), and its solution for absorbed power, Eq. (6-10). It is most important to realize that the cross section σ for almost any process is velocity dependent. Hence, τ_r is velocity dependent; it is directly proportional to $\bar{v}^{-1}$ and also depends on $\bar{v}$ indirectly through σ. At quite moderate rf powers, the driving term $[nq^2/m]\mathbf{E}(t)$ in Eq. (6-9) is capable of affecting a substantial change in the ion kinetic energy and hence in both $\bar{v}$ and σ. So Eq. (6-10) is the correct power supplied to the ions by the marginal oscillator only in the limit of very small E_0. At higher powers, the line shape will not be Lorentzian, since τ_r in Eq. (6-10) will, in fact, depend on $(\omega - \omega_0)$ and E_0. Aside from the question of line shape, the dependence of τ_r on E_0 and $(\omega - \omega_0)$ makes feasible some very clever experiments. A number of these experiments are described by Baldeschwieler [2]. We shall describe one of the many possibilities here.

Suppose ion A^+ and molecule B are admitted to the resonance region and allowed to react. The reaction products after collision will be some ion C^+ and another molecule, D.

$$A^+ + B \rightarrow C^+ + D \tag{6-15}$$

One can just as well observe the ion-cyclotron resonance of C^+ as of the original ion A^+. The resonance of C^+ will occur at a different frequency than A^+ because of its different mass. The C^+ resonance will not be independent of A^+ because by Eq. (6-15) the number of C^+ ions depends on the reaction rate of A^+ with B. That reaction rate is partially under the experimentalist's control by means of the velocity dependence of τ_{A^+}. One can observe with a feeble rf electric field at the ion-cyclotron frequency of species C^+, at ω_2, while irradiating simultaneously with a strong rf electric field at ω near the A^+ resonant frequency, ω_1. While the A^+ ions are heated as ω approaches ω_1, the number of C^+ ions changes and the signal observed by the monitoring oscillator at ω_2 changes. One may chop the power at $\omega = \omega_1$ on and off and detect the

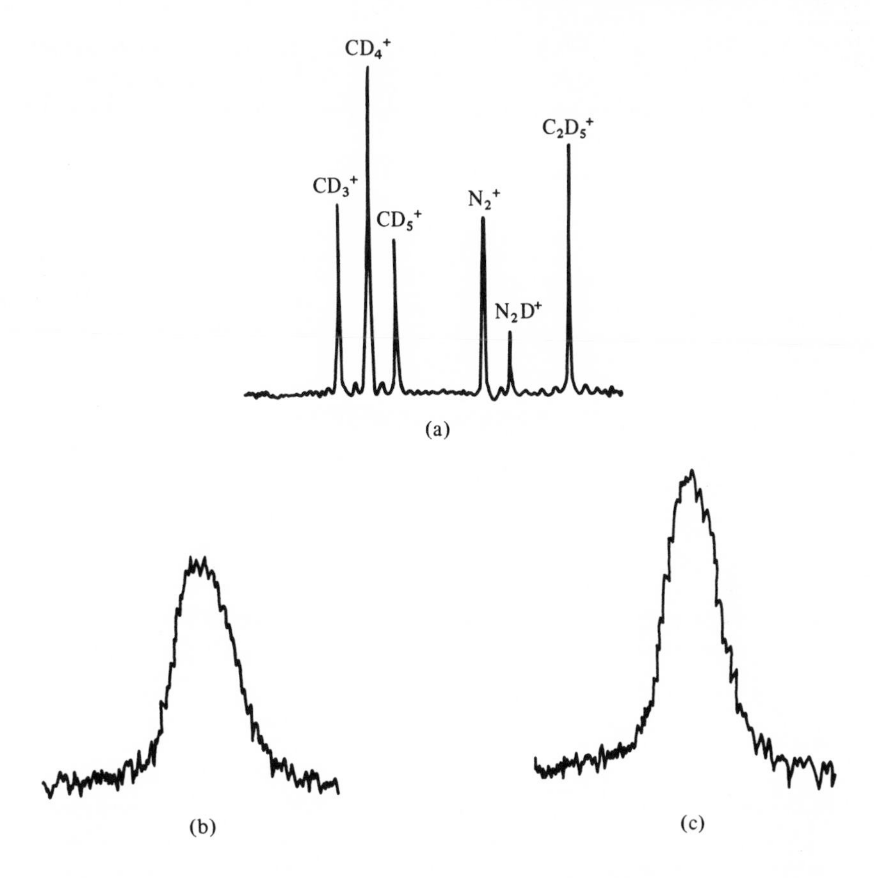

Fig. 6-4 (a) Single-resonance ion-cyclotron resonance spectrum of a mixture of CD_4 and N_2 at a pressure of about 2×10^{-5} Torr. (b) Field-sweep single-resonance spectrum of the N_2D^+ peak with expanded scales. (c) Field-sweep double-resonance spectrum of N_2D^+ resonance with continuous irradiation at the CD_4^+ cyclotron resonance frequency (from Baldeschwieler [2]).

change in the signal due to C^+ synchronously at the chopping frequency. Figure 6-4, reproduced from Baldeschwieler [2], shows the results for the reaction

$$CD_4^+ + N_2 \rightarrow N_2D^+ + CD_3 \tag{6-16}$$

where $A^+ \equiv CD_4$ and $C^+ \equiv N_2D^+$.

Ion-cyclotron resonance experiments seem certain to develop some very elaborate, beautiful, and clever variants. The applications have hardly begun, but commercial apparatus is already available!

Application of cyclotron resonance to solid state physics

Although ion-cyclotron resonance is the newest application of cyclotron resonance, it happens to be the easiest to explain because it involves free ions. The application of cyclotron resonance to the study of electrons in semiconductors and metals is much older, but, perversely, there the phenomenon is more complicated.

To acquire a useful understanding, we must restudy the Lorentz force equation in other language. Our previous discussion has been entirely classical. We can very easily rephrase it in quantum terms, at least for the free particle. We simply use the de Broglie relation between the momentum and the wavelength,

$$p = mv = \frac{h}{\lambda} = \hbar k \tag{6-17}$$

where the wave vector $k = 2\pi/\lambda$ has the same direction as the momentum.

The Lorentz force equation for the free particle in a magnetic field is

$$\hbar \dot{\mathbf{k}} = \frac{e}{c} \mathbf{v} \times \mathbf{B} \tag{6-18}$$

where we have used e as the symbol for charge to indicate that we are thinking about electrons. Equation (6-18) says that the change in $\mathbf{k}$ is normal to both $\mathbf{B}$ and $\mathbf{v}$. We want to be able to view Eq. (6-18) graphically. To do so, we recognize that $\mathbf{B}$ does no work on the electron, so that the motion described by Eq. (6-18) is on a surface of constant energy. The equation of that surface for a free particle is

$$E(k) = \frac{\hbar^2 k^2}{2m} \tag{6-19}$$

The $E(k)$ surface in k space for a free particle is a sphere. The relation between $\mathbf{B}$, $\mathbf{v}$, and the particle trajectory on the constant energy surface of Eq. (6-19) is shown in Fig. 6-5. The particle velocity $\mathbf{v}$ is perpendicular to the spherical surface of constant energy. So, by Eq. (6-18), the trajectory of the electron in the presence of $\mathbf{B}$ is on the circumference of the circle normal to $\mathbf{B}$ which intersects the sphere at the point determined by $\mathbf{v}$ (Fig. 6-5). We can recover one important result in this reciprocal space (k space), the cyclotron period, by computing the time it takes for a

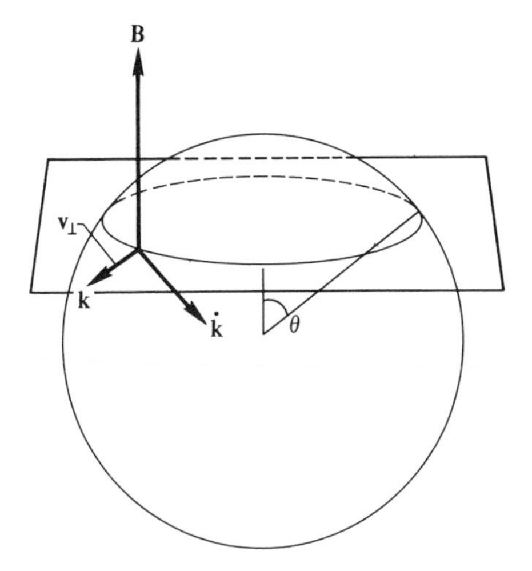

Fig. 6-5 Orbit of an electron on a spherical energy surface in k space in the presence of a magnetic field.

complete circuit. The *rate* at which the wave vector changes is given by Eq. (6-18). If we denote the component of **v** perpendicular to **B** by $v_\perp$, we find for the time of circumnavigation

$$T = \frac{2\pi k_\perp}{|\dot{k}|} = \frac{2\pi k_\perp}{ev_\perp B}\, c\hbar = 2\pi \frac{mc}{eB} = \frac{2\pi}{\omega_c} \tag{6-20}$$

since, by the de Broglie relation, $\hbar\mathbf{k} = m\mathbf{v}$. The thing to emphasize is that ω_c is independent of θ (Fig. 6-5); all particles have the same cyclotron frequency independent of their velocity. The result is true because the constant energy surface in k space is spherical. For reasons that we shall try to explain in the next paragraph, the constant energy surface for electrons in solids is seldom spherical. It is useful to write a more general expression for T that is valid for an arbitrary shape:

$$T = \frac{c\hbar}{eB} \oint \frac{dk}{v_\perp} \tag{6-21}$$

The line integral is taken around the path of the orbit of the surface $E(k)$ in k space. $v_\perp$ is the component of **v** in the plane normal to **B** at the point **k** (Fig. 6-5).

The cyclotron resonance experiment on electrons in solids has, in common with most other measurements of macroscopic and transport quantities,[1] the feature that the measured quantity is response of the electron to an external force. But the *entire* force on the electron produces the *response* that is measured in the experiment. An example that is more specific, but not intended to be representative of any particular experiment, is given by the expression

$$\mathbf{F}_{\text{ext}} + \mathbf{F}_{\text{int}} = m\mathbf{a} \tag{6-22}$$

which gives the *response* (in this case, the acceleration $\mathbf{a}$) to the total force acting on m, both the external force $\mathbf{F}_{\text{ext}}$ and the internal forces $\mathbf{F}_{\text{int}}$, such as the force of the periodic crystal potential. But the experimenter has no control over $\mathbf{F}_{\text{int}}$. To him, the response of the system is to $\mathbf{F}_{\text{ext}}$, and he would prefer to write the Eq. (6-22) first in the form

$$\mathbf{F}_{\text{ext}} = m\left(\mathbf{a} - \frac{\mathbf{F}_{\text{int}}}{m}\right) \tag{6-23}$$

and then somehow factor out the response $\mathbf{a}$ so that one can write

$$\mathbf{F}_{\text{ext}} = \mathbf{a} \cdot (m\mathbf{1} - \mathbf{t}) \tag{6-24}$$

where $\mathbf{1}$ is the unit tensor and $\mathbf{t}$ has the property

$$\mathbf{a} \cdot \mathbf{t} = \mathbf{F}_{\text{int}} \tag{6-25}$$

The quantity in parentheses in Eq. (6-24) is dimensionally a mass, and one could summarize the properties of the response of an electron to only part of total force acting on it by describing the tensor $(\mathbf{m} - \mathbf{t})$.

The quantity so described is in a real solid neither isotropic in general nor is it independent of energy. Most experiments are done in a way that upsets the internal equilibrium of the solid as little as possible, so that the electrons move very nearly on constant energy sufaces. The geometrical shape of a constant energy surface and the way our electron responds to an external force are intimately related. Therefore, the problem of cyclotron resonance must be reexamined in light of these complications.

[1] By transport properties is meant such quantities as electrical and thermal conductivities, and the enormous profusion of galvano-magnetic and galvano-magneto-caloric effects. The cyclotron resonance experiment measures some components of the electrical conductivity tensor as a function of magnetic field. By macroscopic properties other than transport properties are meant specific heat and magnetic and electric susceptibility, for example.

The first cyclotron resonance experiments in solids were done in semiconductors; later, the rather more difficult problem of cyclotron resonance in metals was successfully solved. Not only the experimental problems but also the physics involved in the two types of samples is somewhat different. We begin with semiconductors.

Semiconductors in the ultra pure form have no free electrons (i.e., electrons in the conduction band) on which to do an experiment at 0°K. But in the presence of impurities and nonzero temperatures, a few electrons are raised to the conduction band. The constant energy surfaces on which they move are usually ellipsoids of revolution. That is, the $E(k)$ relation is of the form

$$E(k) = \hbar^2\left(\frac{k_x^{\,2} + k_y^{\,2}}{2m_\perp} + \frac{k_z^{\,2}}{2m_\parallel}\right) \tag{6-26}$$

The shape is conventionally defined in terms of the effective mass tensor, which, in this case, describes an ellipsoid of revolution about the k_z axis, with transverse and longitudinal components $m_\perp$ and $m_\parallel$, respectively. If the magnetic field is applied at an angle θ to the z axis, we obtain Fig. 6-6, which is analogous to Fig. 6-5 for an isotropic energy surface. The rather remarkable thing about an ellipsoidal surface is that all the orbits defined by all the planes which are normal to **B** and which cut the energy

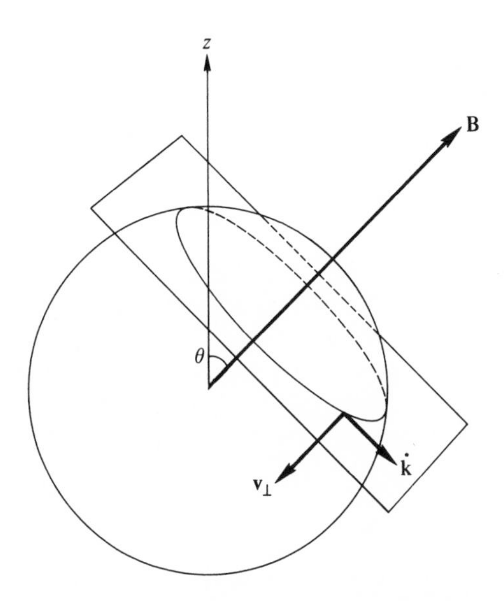

Fig. 6-6 A cyclotron orbit in k space on an ellipsoidal energy surface.

surface have the same cyclotron frequency [3]

$$\omega_c = \frac{eB}{m^*c} \tag{6-27}$$

where

$$\left(\frac{1}{m^*}\right)^2 = \frac{\cos^2\theta}{m_\perp{}^2} + \frac{\sin^2\theta}{m_\perp m_\parallel} \tag{6-28}$$

All the electrons contribute to the signal at the same frequency, given by Eqs. (6-27) and (6-28). The effective mass tensor can be mapped out by measuring the cyclotron resonant field as a function of θ. The cyclotron resonance technique has been most useful in the physics of semiconductors.

The cyclotron resonance problem in metals differs from that in semiconductors both in certain practical details and in the interpretation of the meaning of the quantities measured. Cyclotron resonance in semiconductors and metals is done, as is ion-cyclotron resonance, in fields of a few thousand gauss. Because the mass is that of the electron rather than the ion, the cyclotron frequency is typically in the microwave range 10^{10} to 10^{11} Hz, or more. The high values of ω_c are, in fact, necessary to observe the resonance at all, since the condition necessary for the observation of the resonance, $\omega_c \tau_r \gg 1$, is clearly helped by large ω_c in the face of the short relaxation times τ_r characteristic of solids. In fact, cyclotron resonance was observed first in semiconductors in part because of the early availability of pure samples of semiconductors, with long electron mean free path and hence long τ_r.

Metals differ from semiconductors in two other important respects: the magnitude of the electrical conductivity and the shape of the constant energy surface. The former difference changes the nature of the experimental problem somewhat; the latter changes the interpretation of the results. The important energy of the constant energy surfaces is the Fermi energy, the highest energy for which states are occupied by electrons (see Chapter 5, or Kittel [4]). These surfaces are almost never ellipsoids, as in the case of semiconductors, except for sodium and potassium, which have very nearly spherical Fermi surfaces. In fact, the Fermi surfaces in metals are frequently not even simply connected. That is, rather than forming isolated closed shapes in k space (repeated periodically to reflect the periodicity of the lattice), a given figure joins onto an identical figure in the neighboring unit cell of k space, *ad infinitum*. A hypothetical example is shown in Fig. 6-7. Let the magnetic field direction be exactly along the z direction. The cyclotron orbits on the constant energy surfaces vary from the small neck orbits, dotted in Fig. 6-7b, to the largest, so-called

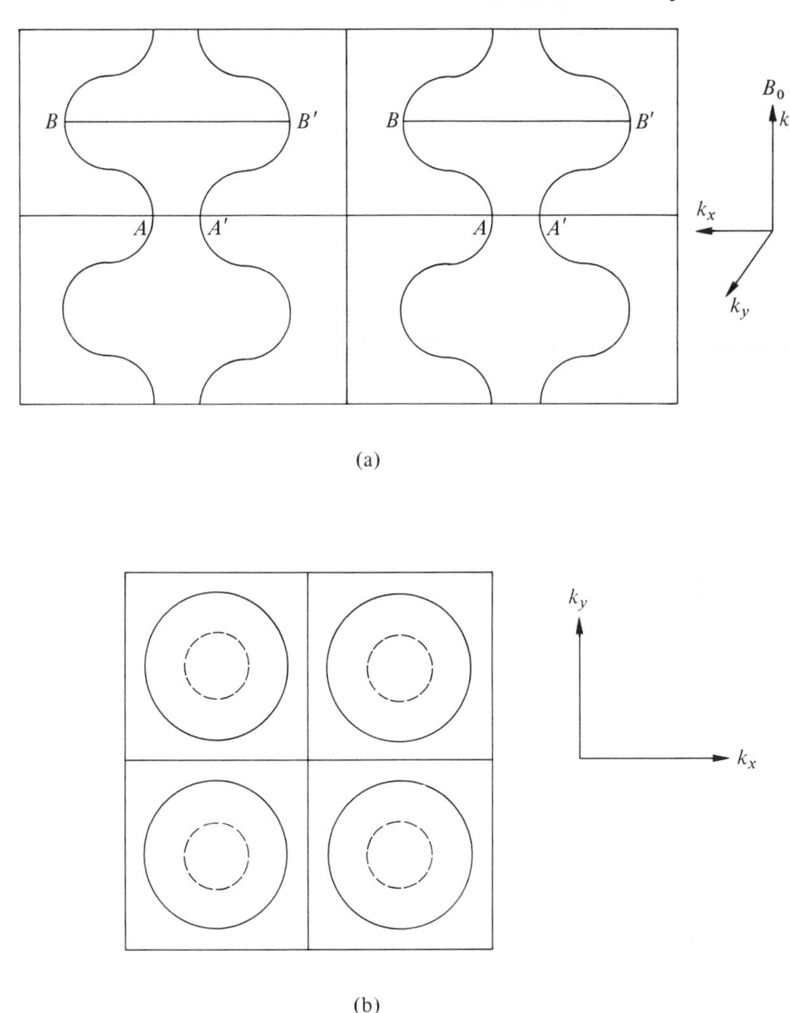

Fig. 6-7 (a) A hypothetical multiply connected Fermi surface showing "neck orbits" at AA' and "belly orbits" at BB'. (b) The cross section in the $k_x k_y$ plane through BB' (solid) and AA' (dotted).

"belly orbits." The cyclotron period is still given by Eq. (6-21), but since the shape of the Fermi surface is no longer ellipsoidal, the period is not the same for each orbit. An analysis of the situation shows that by virtue of the different cyclotron frequencies the signals from the different parts of the Fermi surface cancel each other out except for those orbits which go around slices of the Fermi surface that are extrema. In Fig. 6-7, the extrema are the neck and belly orbits at AA' and BB', when $\mathbf{B}_0$ is pointed along k_z. The *cyclotron* mass, defined by Eq. (6-27), can be written in terms of the Fermi surface as

$$m^* = \frac{\hbar^2}{2\pi} \frac{\partial A}{\partial E} \tag{6-29}$$

where A is the area enclosed by the orbit in the plane normal to **B**. Equation (6-29) can be derived from Eq. (6-22) (see Ziman [5], pp. 250ff.). In metals, then, cyclotron resonance measured the effective mass of these orbits for which $\partial A/\partial E$ is an extremum, either minimum or maximum.

Other orbits are possible in real metals. For example, if a greater density of electrons exists in the metal than the density that leads to the hypothetical Fermi surface of Fig. 6-7, then necks could be formed along the k_x and k_y directions. Fig. 6-8 shows an orbit that would exist if B_0 were along the k_y direction. Such an orbit is a "hole" orbit. The sense of circulation is opposite to the sense for an electron going around full, rather than empty, electron states.

To observe cyclotron resonance in metals, one must solve two problems. One is associated with the technology of pure metals. The condition that must be satisfied to observe cyclotron resonance at all, $\omega_c \tau_r \gg 1$, is satisfied only in pure metals at very low temperatures. The low temperatures are needed to suppress lattice vibrations, which deflect the electron from the cylotron orbit. That stratagem works only if the metal has few

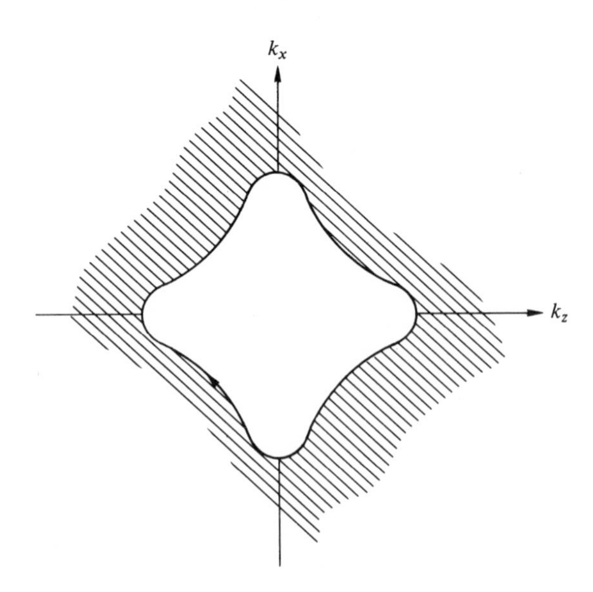

Fig. 6-8 A "hole-orbit" surrounds empty rather than full electron states.

enough impurities that scatter electrons to allow the basic inequality to be satisfied. Typical satisfactory relaxation times are in the vicinity of 10^{-10} sec. Although this may seem a very short time, it should be remembered that a typical velocity of an electron at the Fermi surface is 10^8 cm/sec, so a mean free path is 10^{-2} cm or 100 μ, on the order of a million lattice parameters. The technology that produced metals of the required purity developed somewhat later for metals than for semiconductors, so cyclotron resonance in metals had to wait.

Because of the high conductivity of metals, the rf skin depth is small, as we discussed in Chapter 5. If we use the classical skin depth expression $\delta = c/(2\pi\omega\sigma\mu)^{1/2}$, we would find that the rf field penetrates the metal only a small fraction of a micron, as discussed in Chapter 5. In 3000 G, the cyclotron orbit diameter is

$$d_c = \frac{2v_F}{\omega_c} \simeq 30\mu$$

so the picture of Fig. 6-1, in which the electric field acts on the electron with equal force over its entire orbit, is incorrect. Although our conclusion is right, there was a faulty step along the way. Whenever the formula above for δ predicts a skin depth comparable to or shorter than the mean free path, the assumption under which the skin depth formula was derived is violated. The assumption was that Ohm's law is obeyed:

$$\mathbf{j}(\mathbf{r}) = \sigma\mathbf{E}(\mathbf{r}) \tag{6-30}$$

Equation (6-30) is a *local relation* between the current density $\mathbf{j}(\mathbf{r})$ at the point $\mathbf{r}$ and the field $\mathbf{E}(\mathbf{r})$ at that same point. One can easily see that the current at any point is actually determined in part by the action of the field on the electron since the last electron-lattice collision. If the time lapse between collisions is short compared to the time it takes for the field to change appreciably, and if the field is spatially homogeneous over a mean free path, then the local relation, Ohm's law, is an adequate description of the true state of affairs. But when the classical formula predicts a skin depth of less than 1 μ and the mean free path is 100 μ, the theory is clearly not internally consistent.

The term used to describe the regime of $\Lambda \gtrsim \delta$ is the *anomalous skin effect*. Rather involved calculations show that the rf field penetration into the metal does not fall off exponentially but rather rapidly at first, and then with a long tail extending into the metal as far as Λ. The region of reasonably strong rf field has a thickness on the order of

$$\delta_a = \delta\left(\frac{\Lambda}{\delta}\right)^{1/3} \tag{6-31}$$

Some quick numerical estimates will serve to fix the ideas. The skin depth of, say, copper at 10^{10} Hz and room temperature is about $\delta = 1\ \mu$. A pure sample has a resistance ratio $\rho(300°\text{K})/\rho(4.2°\text{K})$ of about 10^4 ($\rho = \sigma^{-1}$ is the resistivity). The classical skin depth at 4.2°K is 10^{-6} cm $= 0.01\ \mu$. From Eq. (6-35), the anomalous skin depth is about 0.25 μ, a number still much less than the cyclotron orbit diameter. Actually, Eq. (6-31) does not express the facts in the presence of a static field B_0 large enough that $\omega_c \tau_r \gg 1$. Under those circumstances, the field penetration is complicated, geometry dependent, and not necessarily monotonic. Nevertheless, our original conclusion remains correct: we cannot accelerate the electron during its entire orbit with the rf electric field because of the small penetration of the field into the metal sample. Cyclotron resonance in metals is still possible if the proper geometrical arrangement of sample, static, and rf fields is chosen. The most widely used arrangement was suggested by the Russian physicists Azbel' and Kaner [6]. It actually bears some relationship to the original cyclotron of Lawrence (Fig. 6-1), except the rf electric field acts on the electron at most once per cycle rather than twice. The geometry and an electron cyclotron orbit are shown in Fig. 6-9.

The rf electric field tangential to the surface accelerates those electrons having cylcotron orbits that graze the metal surface. An rf surface current is produced, and power is taken from the rf generator proportional to $\mathbf{j}_{\text{rf}} \cdot \mathbf{E}_{\text{rf}}$. One additional feature of this geometry is not shared by the other cyclotron resonance experiments we have discussed. The rf field and the rf surface currents are in phase whenever the rf field has completed

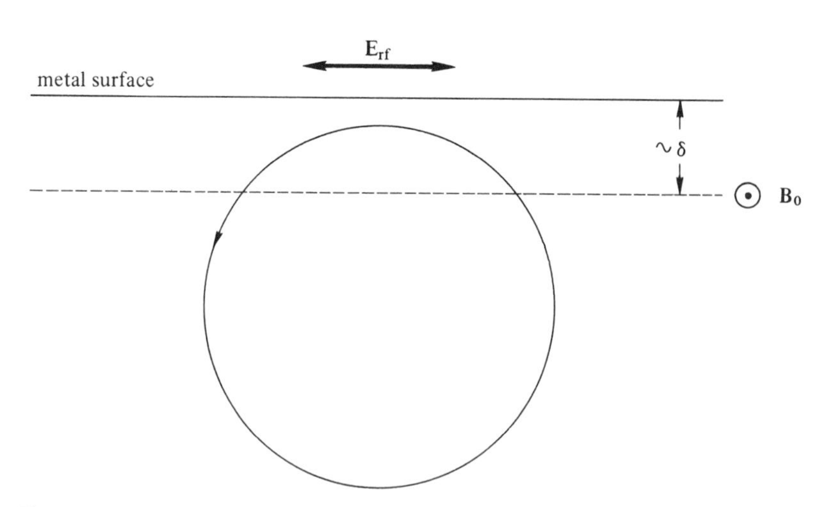

Fig. 6-9 Azbel'–Kaner cyclotron resonance geometry. The static magnetic field is perpendicular to the page, and the cyclotron radius is greater than the rf field penetration depth.

n periods for every cyclotron period, where $n = 1, 2, 3, \ldots$. Power is absorbed for periods $nT = T_c$, or $n\omega_c = \omega$—that is, at fields B_0, for constant rf frequency ω, given by

$$\frac{1}{B_0(n)} = \frac{ne}{\omega m^* c} \tag{6-32}$$

Figure 6-10a shows Azbel'–Kaner type cyclotron resonance in potassium. If the fields at which the power absorption is a maximum are identified, and their reciprocals plotted versus integers, the result is a straight line. The slope of the line is proportional to the inverse of the cyclotron mass m^* for the electrons on an extremal orbit.

Cyclotron resonance in metals has been useful in determining something about the Fermi surface through the cyclotron mass, as given in general by Eq. (6-29). The shapes of the Fermi surfaces in a large number of metals have been determined by cyclotron resonance and a variety of other experiments that measure other properties of the Fermi surface, such as the de Haas–van Alphen effect, magnetoacoustic attenuation, and others. (For a review and references, see the little book by Ziman [7].)

One other experiment very closely related to cyclotron resonance goes by the name "size effect." We shall conclude our discussion of cyclotron resonance with a brief description. Imagine that the metal sample in Fig. 6-9 is terminated by a flat surface accurately parallel to the upper surface at a depth below the top surface equal to the cyclotron orbit diameter. The lower surface ought then to have induced surface currents 180° out of phase with the upper surface currents, and these currents can, in principle, be detected by the electromagnetic field they produce on the lower side of Fig. 6-9. Actually, the measurements are more easily done by wrapping an rf coil around a slab-shaped sample (Fig. 6-11) and allowing the coil to be part of the tank circuit of a nuclear magnetic resonance marginal oscillator circuit oscillating at any frequency. The experiment then measures the surface impedance of the sample as a function of magnetic field. Anomalies in the surface impedance appear at fields that cause the cyclotron orbit diameter to be t/n, where t is the sample thickness and n is an integer:

$$\frac{t}{n} = \frac{4\pi v_F}{\omega_c} = \frac{4\pi}{eB_0} m^* v_F \tag{6-33}$$

Equation (6-33) is valid for a spherical Fermi surface, but gives the correct result for arbitrary Fermi surface shape if Eq. (6-33) is correctly interpreted. The measured quantity is the Fermi momentum $m^* v_F = \hbar k_F$, which for a nonspherical surface is to be interpreted as the maximum

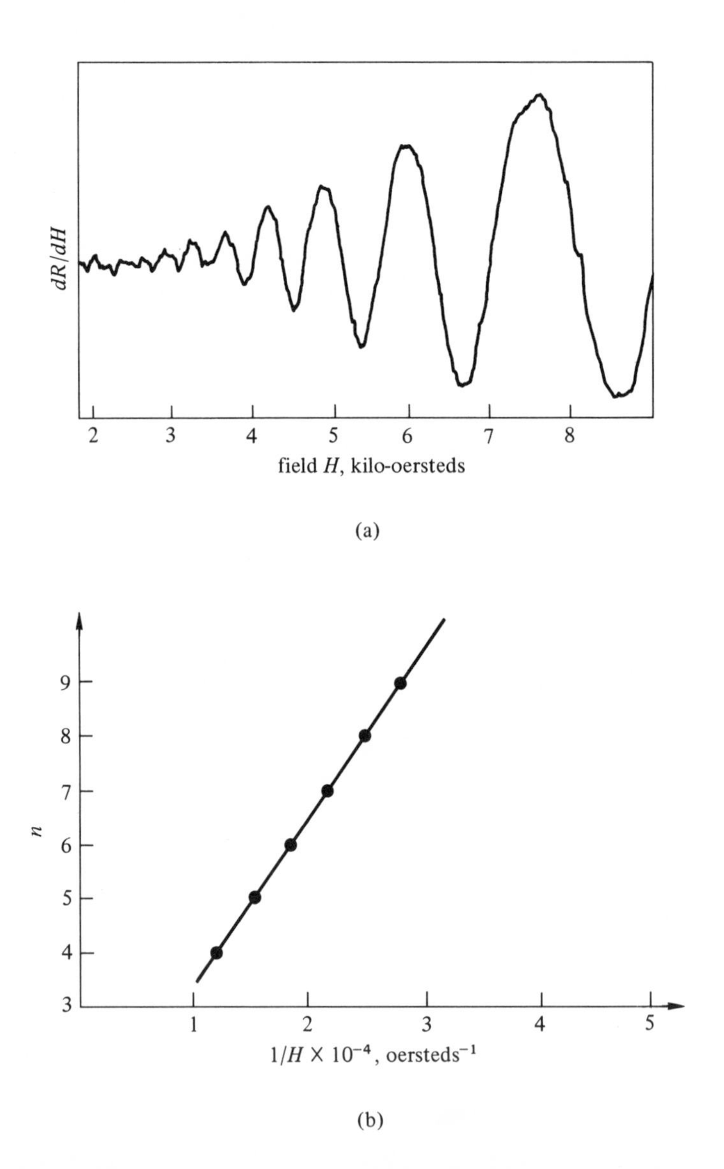

Fig. 6-10 Azbel'–Kaner cyclotron resonance in potassium. (a) The derivative of the real part of the microwave surface impedance at 66.2 GHz versus the magnetic field [after C. C. Grimes and A. F. Kip, *Phys. Rev.* **132**, 1991 (1963)]. (b) Reciprocal of the position of absorption maxima in oersteds^{-1} versus integers, from (a). Note that only subharmonics with $n \geq 4$ are seen in the available fields.

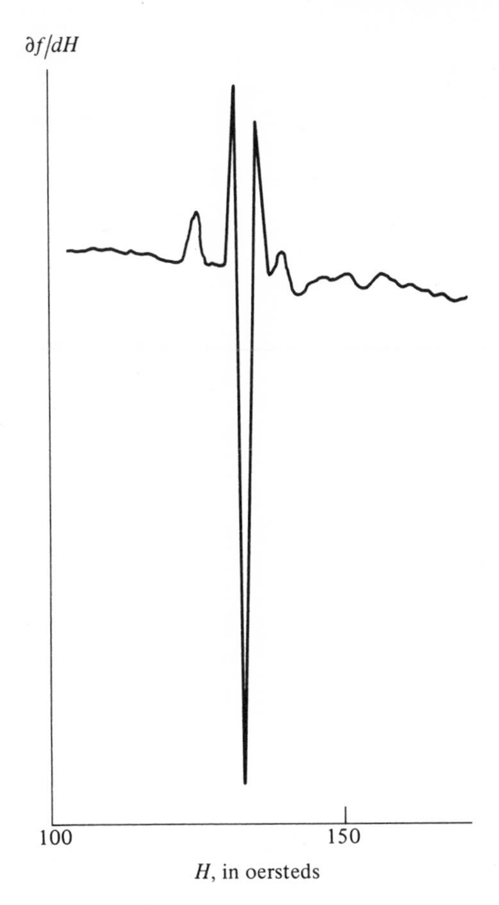

Fig. 6-11 Radiofrequency size effect experiment in a tin sample 0.54 mm thick at 2.4 MHz and 3.75°K. Plotted is the derivative with respect to H of the frequency of a marginal oscillator the tank coil of which is wrapped around the sample. The integer n in Eq. (6-33) is $n = 1$ [after V. K. Gantmakher, *Sov. Phys.—JETP* **16**, 247 (1963)].

Fermi momentum in the plane perpendicular to $\mathbf{B}_0$. The measurement requires only knowledge of t and $\mathbf{B}_0$. The integer n occurs because for cyclotron diameters an integral fraction of t, internal current sheets are formed that, in turn, induce current sheets another cyclotron diameter lower into the metal. Surface impedance anomalies appear whenever an integral number of current sheets just fit into the sample. Figure 6-11 shows an example for tin. Notice that nothing specific was said about the frequency to which the marginal oscillator is to be tuned. In fact, the surface impedance anomalies appear whenever Eq. (6-33) is satisfied at any frequency. The anomalies are sharper if the rf skin depth is small compared to t, but otherwise the oscillator frequency does not matter. The size effect experiment is not a resonance experiment at all; or rather, it is not a temporal resonance, but a spatial resonance, so perhaps it is time to take the cue and move on.

6-2. OPTICAL PUMPING

We mentioned at the beginning of this book that it is convenient to classify magnetic resonance experiments by how the magnetization is produced and how it is detected. Most of the book has been concerned with the equilibrium magnetization produced by an external magnetic field acting on a sample at some unique temperature, and with the direct detection of this magnetization by appropriate pick-up coils or by measurement of energy absorption. The initial magnetic resonance experiments in atomic beams are examples in which both the polarization and the detection of a change in polarization are accomplished by some other means. Optical pumping is another such technique; in its experimentally simplest form, the detection of the resonance is done by monitoring the producer of the polarization, a light beam transmitted through a gaseous sample. In addition to the precise information about atomic structure obtained from optical pumping experiments, the technique has given us a number of interesting devices that have wide application in a variety of fields. Two examples are magnetometers used in geophysics, oceanography, and archaeology,[2] and very stable secondary frequency standards used in space technology.

A wide variety of experiments may accurately be included under the heading of optical pumping. A survey of them and reprints of many of the important original papers in the field are to be found in the monograph and reprint volume by Bernheim [8]. We shall attempt no such survey but restrict our discussion to the most common and easily performed experiments.

There are two rather separate facets to an optical pumping experiment on, say, rubidium vapor. One is the general idea of the experiment, and the other is the atomic physics of the rubidium atom—the nature of the ground and excited states, including the hyperfine structure, and even, for rubidium anyway, isotope effects. The atomic structure part is a little complicated, so rather than try to explain an optical pumping experiment on a real system, we shall separate the facets and analyze first a simple hypothetical atomic system. Then we shall show how the basic ideas are applied to a real atomic system.

[2] The ancient, almost legendary city of Sybaris was recently uncovered with the aid of a cesium vapor magnetometer.

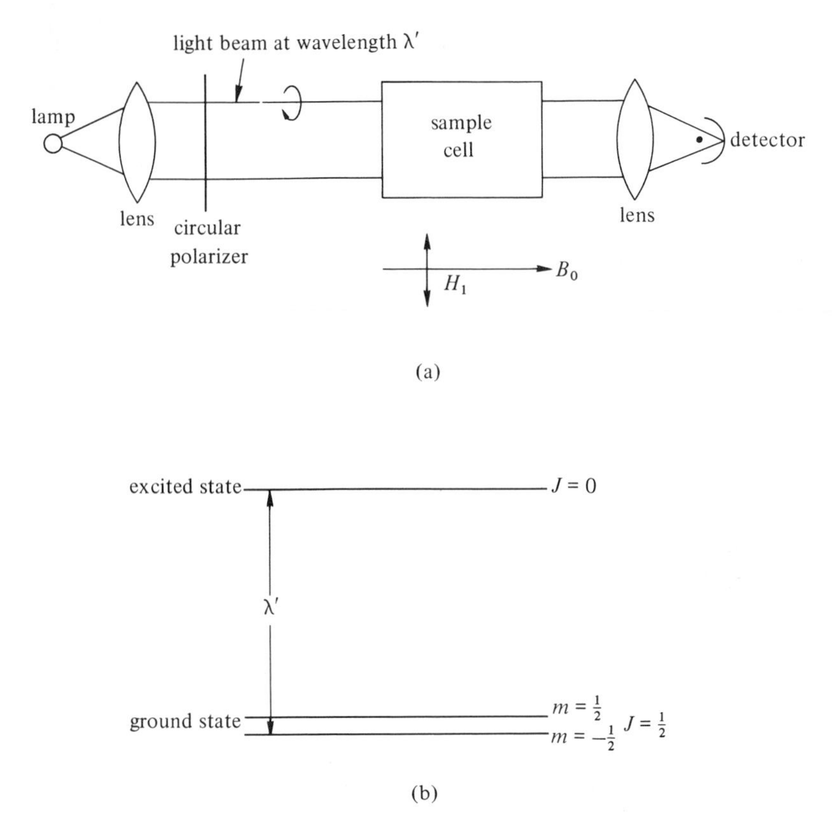

Fig. 6-12 (a) Arrangement of basic components of optical pumping apparatus. (b) Energy-level diagram of a hypothetical atomic system for a sample. The ground state is a Kramer's doublet split by the small field B_0.

Figure 6-12 shows the basic arrangement of the experiment we shall analyze. The lamp is an atomic discharge from the same atoms present as a gas in the sample cell. Assume that most of the radiation from the lamp is at the wavelength λ', corresponding to the transition from the excited state of Fig. 6-12b to the doubly degenerate ground state whose degeneracy is lifted by the field B_0. The wavelength λ' is usually in the ultraviolet, visible, or near infrared, and the energy spread of the spectral line is broader than the energy separation between the Zeeman split levels of the ground state. We further assume the ground state splitting to be so small compared to kT that the ground state polarization from the Boltzmann factor is negligible—both states are equally occupied. The beam from the lamp is circularly polarized before it strikes the sample, which means that each incident photon carries one unit of angular momentum

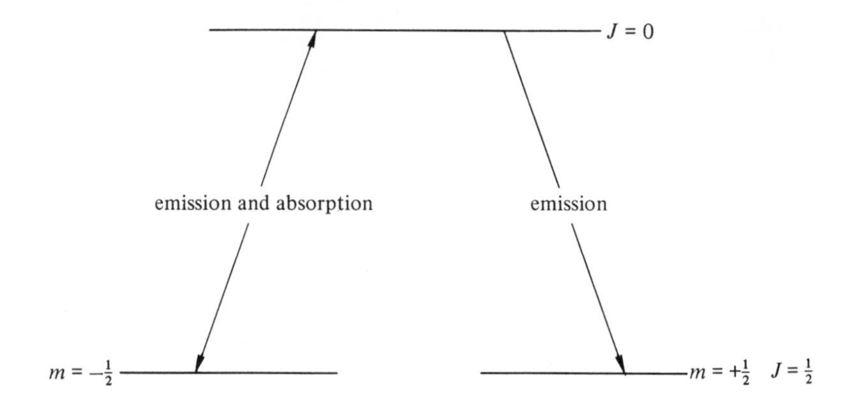

Fig. 6-13 Allowed absorption and emission transitions of hypothetical sample atom. Incident photons have $+\hbar$ angular momentum only.

either parallel or antiparallel to the propagation direction, the z direction.[3] We further assume that the single excited state is nondegenerate and has angular momentum properties such that the photon emitted in a transition to the $m = -\frac{1}{2}$ ground state has $+\hbar$ angular momentum, and the photon emitted in the transition for the $-\frac{1}{2}$ state has a $-\hbar$ angular momentum. Thus the selection rules for absorption in the sample cell are as follows: the $-\frac{1}{2}$ state can absorb only a $+\hbar$ photon, the $+\frac{1}{2}$ state can absorb only a $-\hbar$ photon. Let the cross section for this process be σ. Once a sample atom is excited, it reemits a photon to either of the ground states with equal probability, but the emission is spatially isotropic. Thus, there is very small probability that the photon is emitted into the small solid angle that would allow it to get to the photocell (detector).

These complicated requirements only state in words the essential features of the selection rules that will govern the absorption and emission of resonance radiation in a real atomic system. If the circular polarizer in Fig. 6-12 allows only photons of $+\hbar$ to reach the sample, then the transitions of Fig. 6-13 are allowed for the atoms in the sample. We see

[3] See Feynman [9], vol. 3, Chap. 19, for a discussion of the angular momentum properties of the photon. Be prepared for the old confusion about the convention defining right- and left-hand circularly polarized light. For a photon that has $+\hbar$ angular momentum relative to the propagation direction, the electric vector circulates in the direction indicated by the fingers of the right hand while the thumb points in the propagation direction. Feynman calls this RHC polarized light. The electric vector is rotating counterclockwise as seen by an observer looking at the beam while facing the source. The old convention of classical optics defines this sense of rotation to be left-hand circularly polarized (LHC). We shall avoid the confusion by always referring to the direction of the angular momentum of the photon, since we shall only be referring to circularly polarized light.

immediately that the $-\frac{1}{2}$ state has both absorption and emission processes connecting it to the excited state, but the $+\frac{1}{2}$ state is only fed by the excited state—no absorption from it is allowed by the selection rules. The result is that the population of the $+\frac{1}{2}$ state grows until, if there are no competing processes, all the atoms will be in the $+\frac{1}{2}$ state. At this point, the gas will be completely transparent to the pumping light, since there are no atoms in the $-\frac{1}{2}$ state to absorb the incident light. We see further that if we repopulate the $-\frac{1}{2}$ state by, for example, doing a magnetic resonance experiment on the ground state, the intensity of the light reaching the detector will drop. Monitoring the intensity of the transmitted light thus provides a very simple way of detecting the magnetic resonance on the ground state.

It is instructive to rephrase the preceding description in somewhat more general and macroscopic terms, so that we can begin to divorce ourselves from this particular three-level model. The incident beam carries angular momentum in the $+z$ direction. The beam is partially absorbed, and the absorber acquires some of that angular momentum. By means of that great friend of magnetic resonance, the Wigner–Eckart theorem, the gas acquires a magnetic moment, and it gives the experimenter a handle on the angular momentum. Since the atoms are in a gas, which implies a short correlation time, it should not be unreasonable to describe the interaction of the magnetic moment with the static and rf magnetic field by the Bloch equations. These equations would also contain, of course, relaxation terms between the two ground state levels. The additional feature of the optical pumping experiment is the interaction between the pumping light and the sample. We shall proceed now to derive the term that must be added to the Bloch equations to describe the interaction with the pumping light.

Consider a volume element of the sample of unit cross section and thickness in the propagation direction dz. Let $n(z)$ be the incident photon flux: the number of incident circularly polarized photons at wavelength λ per second per unit area at z. Let σ be the cross section for absorption of a photon by an atom in the $-\frac{1}{2}$ state. The number of photons lost from the beam in thickness dz is

$$dn = -N_{-1/2}(z)n(z)\sigma\, dz \tag{6-34}$$

where $N_{-1/2}(z)$ is the number of atoms per unit volume in the ground state. The rate at which atoms are lost from the $-\frac{1}{2}$ state due to the pumping process is related to the annihilation rate of the photons dn by

$$\frac{1}{2}\, dn = \left(\frac{\partial N_{-1/2}}{\partial t}\right)_{\text{rad}} dz \tag{6-35}$$

where the factor of $\frac{1}{2}$ occurs because we assume the reemission rate to each of the ground states from the excited state is equal, so that half of the absorbed photons do not cause a net change in $N_{-1/2}$. The partial derivative and the subscript "rad" are to indicate that there are other ways that $N_{-1/2}$ can change. Note that we implicitly regard the absorption-reemission process to be instantaneous. Equations (6-34) and (6-35) combine to yield

$$\frac{dN_{-1/2}}{dt} = -\frac{n(z)}{2}\sigma N_{-1/2}(z) \tag{6-36}$$

We can convert this equation to an equation for the z component of the sample magnetization by use of the following definitions and relations. The z component of the angular momentum of the gas is

$$J_z = \frac{\hbar}{2}(N_{1/2} - N_{-1/2}) \tag{6-37}$$

and the magnetic moment is

$$M_z = \gamma J_z = \frac{\gamma\hbar}{2}(N_{1/2} - N_{-1/2}) = \frac{\gamma\hbar}{2}\mathcal{N} \tag{6-38}$$

where

$$\mathcal{N} = (N_{1/2} - N_{-1/2}) \tag{6-39}$$

and γ is the appropriate gyromagnetic ratio. Since we have Eq. (6-36) for $N_{-1/2}$, we want to write $N_{-1/2}$ in terms of $\mathcal{N}$. Let N_0 be the total number density of atoms in the sample.

$$N_0 = N_{1/2} + N_{-1/2} \tag{6-40}$$

Hence

$$N_{-1/2} = \frac{N_0 - \mathcal{N}}{2} \tag{6-41}$$

From Eqs. (6-36), (6-38), and (6-41), we obtain

$$\left(\frac{\partial M_z}{\partial t}\right)_{\text{rad}} = \gamma\hbar\frac{n(z)}{2}\sigma\frac{1}{2}(N_0 - \mathcal{N}) \tag{6-42}$$

or

$$\left(\frac{\partial M_z}{\partial t}\right)_{\text{rad}} = n(z)\frac{\sigma}{2}(M_0 - M_z) \tag{6-43}$$

or

$$\left(\frac{\partial M_z}{\partial t}\right)_{\text{rad}} = \frac{M_0 - M_z}{\tau_p} \tag{6-44}$$

where $M_0 = \gamma\hbar N_0/2$ is the maximum magnetization, achieved when all the atoms are in the $+\frac{1}{2}$ state. The pumping time, $\tau_p = 2[n(z)\sigma]^{-1}$, is determined by the light intensity and, of course, σ. In the usual unit of light intensity, which is incident power per unit area, we write

$$I(z) = \hbar\omega' n(z) \tag{6-45}$$

where $\omega' = 2\pi c/\lambda'$. The total rate of change of the magnetization is then

$$\frac{\partial \mathbf{M}(z)}{\partial t} = \gamma \mathbf{M} \times \mathbf{B} - \frac{(\hat{\mathbf{i}}M_x + \hat{\mathbf{j}}M_y)}{T_2} - \frac{M_z\hat{\mathbf{k}}}{T_1} + \frac{M_0 - M_z}{\tau_p(z)}\hat{\mathbf{k}} \tag{6-46}$$

where we have assumed different transverse and longitudinal relaxation times, T_1 and T_2, and that the equilibrium M_z without pumping is negligible. Equation (6-46) is valid only for the thin slab of sample at z, thickness dz. Notice $\tau_p(z)$, which depends on $I(z)$, has been written explicitly as a function of z. We assume that diffusion of sample atoms from one part of the cell to another can be neglected.

Our job is not complete until we display the other of the pair of coupled equations, namely, the equation for $I(z)$, particularly since the experiment depends on measuring $I(l)$, where l is the length of the sample cell. From Eqs. (6-35), (6-38), (6-41), and (6-45), we can write

$$\frac{dI}{dz} = -\frac{\omega'}{\gamma}\left(\frac{\partial M_z}{\partial t}\right)_{\text{rad}} \tag{6-47}$$

or, in terms of τ_p^{-1},

$$\frac{d\tau_p^{-1}}{dz} = -\frac{\sigma}{2\gamma\hbar}\left(\frac{\partial M_z}{\partial t}\right)_{\text{rad}} \tag{6-48}$$

[Good exercise: Check Eq. (6-48) dimensionally.] Equations (6-46) and (6-47) or (6-48) are the coupled equations describing the interacting systems of the light and the atomic system. The formal solution for the measured intensity $I(l)$ may be written

$$I(l) - I(0) = \int_0^l dz \frac{dI}{dz} = -\frac{\omega'}{\gamma} \int_0^l \left(\frac{\partial M_z}{\partial t}\right)_{\text{rad}} dz \tag{6-49}$$

where $(\partial M_z/\partial t)_{\text{rad}}$ is given by Eq. (6-44).

A general solution of Eq. (6-49) is quite difficult; it is nonlinear, for one thing, but we can learn a little about the characteristics of the experiment by taking a simple case and letting intuition guide us in more complicated cases. Consider the steady state situation, $dM_z/dt = 0$, under the circumstance that the sample is optically thin. In practice, that means ignoring the z dependence of M_z and τ_p in Eq. (6-46). That equation may then be written

$$\frac{d\mathbf{M}}{dt} = \gamma \mathbf{M} \times \mathbf{B} - \frac{(\hat{\mathbf{i}} M_x + \hat{\mathbf{j}} M_y)}{T_2} + \frac{(M_0' - M_z)}{T_1'} \hat{\mathbf{k}} \tag{6-50}$$

where

$$M_0' = \frac{M_0 T_1}{T_1 + \tau_p} \tag{6-51}$$

and

$$\frac{1}{T_1'} = \frac{1}{T_1} + \frac{1}{\tau_p} \tag{6-52}$$

The interpretation of Eq. (6-50) is particularly simple. It is formally identical to the Bloch equations with unequal transverse (T_2) and longitudinal (T_1') relaxation times. The inverse of T_1', as seen in Eq. (6-52), is a pseudo spin-lattice relaxation rate determined by the parallel processes of pumping and true spin-lattice relaxation. The magnetization toward which the z component of M relaxes at the rate $T_1'^{-1}$ is M_0', Eq. (6-51), which is seen to be always less than M_0. M_0' is established in the competition between the pumping light, which pumps toward M_0', and the spin-lattice relaxation, which relaxes toward $M_z = 0$. In the presence of an rf field transverse to B_0, the steady state solution of M_z is, from Eq. (2-45),

$$M_z(\omega) = M_0' \left[\frac{1 + (\omega - \omega_0)^2 {T_2}^2}{1 + (\omega - \omega_0)^2 {T_2}^2 + \gamma^2 {H_1}^2 T_2 T_1'}\right] \tag{6-53}$$

Of course, the signal is determined by the difference in the transmitted intensity in the presence and absence of the radio frequency:

$$S = \frac{dI_z}{dz}\,\Delta z = \left(\frac{M_0 - M_z}{\tau_p}\right)\left(\frac{-\omega'}{\gamma}\right)\Delta z \tag{6-54}$$

where Δz is the sample thickness, and M_z is to be given by Eq. (6-53).

We can exploit our solution of the thin sample case to discuss other experiments. For example, one may measure T_1 by doing a transient experiment without using an rf field at all. After the equilibrium polarization M_0' is reached, the magnetic field $\mathbf{B}_0$ is suddenly reversed. This has the effect of reversing the roles of the $\frac{1}{2}$ and $-\frac{1}{2}$ states in our model system. The action of the pumping light is now to depopulate the $+\frac{1}{2}$ states and populate the $-\frac{1}{2}$ level until the new equilibrium polarization is established, with M_0' pointing in the z direction. Immediately after the field reversal, the gas becomes more opaque and it becomes more transparent again with the time constant T_1' of Eq. (6-52). T_1' depends on τ_p as well as the desired T_1, so the experiment must be done as a function of pumping light intensity and the results extrapolated to zero intensity.

Another experiment that may be analyzed in terms of the Bloch equations involves the use of two light beams. The second one, from a lamp identical to the first, is directed transverse to the magnetic field—say, in the x direction. Now pumping terms must be derived to obtain the appropriate equation analogous to Eq. (6-52) for the thin sample case. The major change is now that T_2 must include the effect of the x direction light. [It already contains $a(\tau_p^{-1})_z$ from the z component pumping light, a result not mentioned previously because we were not primarily concerned with T_1. The dependence of T_2 on the z direction light arises from the lifetime limiting effect of the pumping action on the $-\frac{1}{2}$ state.] Similarly, T_1 will depend on $(\tau_p^{-1})_x$. When a magnetic resonance experiment is done, a transverse magnetization is induced at the Larmor frequency. The effect of the transverse magnetization will be to modulate the transmitted intensity of the x light at the Larmor frequency. This result was demonstrated by Bell and Bloom ([8]; see reprints 8 and 9 in that volume), who further pointed out that one can dispense with two beams and operate with one light beam inclined at some angle, say 45°, with respect to the magnetic field direction. The magnetic resonance will modulate that beam at frequencies near zero, γB_0, $2\gamma B_0$, and so on.

The Larmor frequency modulation of the beam forms the basis of a very sensitive magnetic field measuring device. Since the gyromagnetic ratio γ is determined mainly by the electron gyromagnetic ratio ($\nu_L = 750$ kHz/G for ^{87}Rb, as we shall see), modulation frequencies of several hundred kilocycles are obtained for the earth's magnetic field ($\frac{1}{2}$ G).

Since T_1 and T_2 may be made as long as several tenths of seconds, the accuracy of such an optical pumping device operated as a magnetometer is a fraction of "gamma" (1 gamma $= 10^{-5}$ G). Such sensitivity is useful both for geological studies (including even remunerative ventures, such as oil prospecting) and space studies.

At this point it is well to face the complexities of the real world, and apply the preceding discussion to the atomic systems very commonly used in optical pumping, the alkali atoms. The discussion need not be directed toward any particular alkali, since the important features of the atomic structure are the same for each. The nuclear spin plays an important role, however. Alkali atomic nuclei are most frequently spin $\frac{3}{2}$ (^{7}Li, ^{23}Na, ^{39}K, ^{41}K, ^{85}Rb), with $I = \frac{5}{2}$ (^{87}Rb) and $I = \frac{7}{2}$ (^{133}Cs) the only important exceptions. Therefore we shall use spin $\frac{3}{2}$ throughout the discussion when specific examples are needed.

The ground electronic state of an alkali is $n\,^2S_{1/2}$, from the configuration ns ($n = 2, 3, 4, 5, 6$). The lowest excited electronic state comes from the excitation of the valence electron to the np configuration, giving rise to the famous alkali doublet states $n\,^2P_{3/2,\,1/2}$. The resonance radiation supplying the pumping light comes from the $n\,^2P \rightarrow n\,^2S$ transition. The wavelengths for the alkalis are in the visible and near infrared (5890 Å and 5896 Å for Na). In the ground state, the hyperfine interaction between the electron's intrinsic magnetic moment and the nuclear magnetic moment was shown, originally by Fermi, to have the form

$$\mathscr{H}_{\text{hfs}} = \frac{8\pi}{3} \mu_n \mu_e \hbar^2 \mathbf{I} \cdot \mathbf{S}\, \delta(r) \tag{6-55}$$

at least for s electrons. Taking the expectation values of $\mathscr{H}_{\text{hfs}}$ over just the spatial coordinates leaves the familiar form involving just the spin operators:

$$\mathscr{H}_{\text{spin}} = \frac{8\pi}{3} \mu_e \mu_n \hbar^2 \, |\psi_{ns}(0)|^2 \mathbf{I} \cdot \mathbf{S} \tag{6-56}$$

where $|\psi_{ns}(0)|^2$ is the probability density at the nucleus for the ns configuration. The hfs in the excited state is substantially smaller than in the ground state, since it arises from the dipole-dipole interaction averaged over the p wavefunction of the np configuration. Equation (6-55) gives a contribution only for s states, since $\psi(0) = 0$ for all but s states. The consequence is that the spectral line may be produced in a properly constructed lamp to show that partially resolved hfs is unresolved. However, the excited state hfs does have one important consequence, as we shall see.

The spin energy levels from Eq. (6-56) may be obtained quite easily from the vector model of the coupling of the angular momenta **S** and **I**. In zero external magnetic field, the total angular momentum is

$$\mathbf{F} = \mathbf{S} + \mathbf{I} \tag{6-57}$$

which leads to the angular momentum quantum numbers of the ground state $F = I \pm \frac{1}{2}$, or $F = 2, 1$ for $I = \frac{3}{2}$. The energy splitting between the two hfs levels is easily found in the vector model by the standard trick of computing $\mathbf{I} \cdot \mathbf{S}$ in terms of I, S, and F:

$$\mathbf{I} \cdot \mathbf{S} = \frac{F^2 - I^2 - S^2}{2} \tag{6-58}$$

where F^2, I^2, and S^2 are, of course, to be evaluated as $F(F+1)$, $I(I+1)$, and $\frac{3}{4}$, respectively. The ground state splitting is then

$$W(I + \tfrac{1}{2}) - W(I - \tfrac{1}{2}) = hAF \tag{6-59}$$

where A is the coefficient of $\mathbf{I} \cdot \mathbf{S}$ in Eq. (6-56), expressed for convenience in units of frequency. The frequencies, A, for the alkalis range from 10^9 to 10^{10} Hz: microwave frequencies. The ground states are $2F + 1$ degenerate: five- and three-fold degenerate for $I = \frac{3}{2}$. The degeneracy may be removed by a magnetic field. If the field is weak enough, the Zeeman effect approximation may be used. The energy levels are given approximately by

$$W(F, M_F) = \frac{hA}{2}\,[F(F+1) - I(I+1) - S(S+1)] - B_0 M_F \mu_0 g_F \tag{6-60}$$

where

$$g_F = -\frac{g_e}{2}\left[\frac{F(F+1) + S(S+1) - I(I+1)}{F(F+1)}\right]$$

is the Landé g factor or spectroscopic splitting factor.

We have ignored an additional small term in the energy, proportional to μ_n. All this information is summarized in Fig. 6-14, where in (a), the ground state and excited state hfs appear with no attempt to show the Zeeman splitting. The m_F states of both ground and excited states are

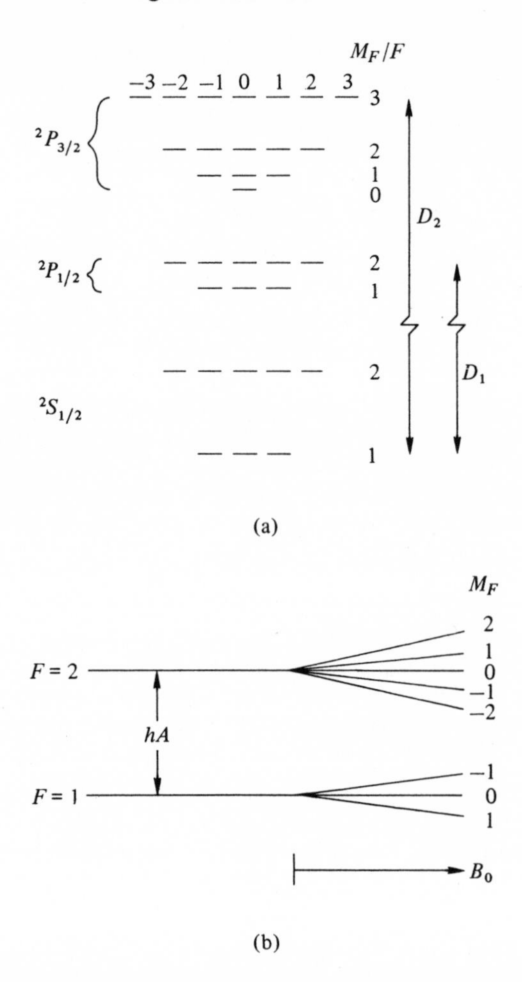

Fig. 6-14 (a) First excited state and ground state hfs for a nuclear spin $\frac{3}{2}$ alkali atom showing order of F levels and conventional labeling of the optical transitions. (b) Ground state Zeeman effect. Neither figure is drawn to scale.

shown separated in the traditional way, so that the overall effect of $\Delta M_F = +1$ transitions can be easily visualized. Figure 6-14b shows the Zeeman effect of the ground states. The field producing the Zeeman effect must be only a few gauss, or deviations from Eq. (6-60) due to the onset of the Paschen–Bach effect can be detected. In Fig. 6-14a, the longer wavelength D_1 and shorter wavelength D_2 transitions are indicated. The figure is badly out of scale: The optical transitions are several electron volts, say, more than 10^4 cm^{-1}; the separation between the $^2P_{3/2}$ and $^2P_{1/2}$ levels is tens to hundreds of cm^{-1}, the ground state $F = 2$ to $F = 1$

splitting is less than 0.3 cm^{-1}, and the excited state hfs splittings are at least an order of magnitude less. In the ground state Zeeman effect, note that magnitudes of the g factors for $F = 2$ and $F = 1$ are the same, but they are of opposite sign. Thus, the M_F levels of the $F = 1$ state are inverted relative to those of the $F = 2$ state.

The experimental arrangement for optical pumping is the same as Fig. 6-12. The sample bulb is filled, usually, to a few millimeters pressure with a pure "buffer" gas, an inert or nonreactive gas, such as a noble gas or N_2. The alkali atoms are evaporated from a small amount of the metal that is also placed in the bulb. The number density of sample atoms can be varied by varying the temperature of the bulb upward from the melting temperature of the metal—about room temperature for Cs to about 100°C for Na. The buffer gas is not essential; it serves to prevent diffusion of sample atoms to the sample cell wall, where relaxation can take place. Wall coatings exist that prevent wall relaxation, however, so successful experiments do not *require* the buffer gas. The pumping light is obtained from a similar bulb, usually smaller, in which an rf discharge is maintained. The output light power from the bulb is overwhelmingly in the D_1 and D_2 resonance radiation lines.

Circularly polarized photons of the D_1 wavelength incident on the sample cause transitions from the ground state to the $^2P_{1/2}$ excited state. The electric dipole selection rules for LHP photons ($+\hbar$ angular momentum) that govern the absorption of these photons are

$$\begin{aligned} \Delta l &= +1 \\ \Delta M_F &= +1 \\ \Delta F &= 0, \pm 1 \end{aligned} \tag{6-61}$$

Thus, the D_1 light pumps atoms from the F, M_F ground state to the F or $F \pm 1$, $M_F + 1$ state of the $^2P_{1/2}$ levels. Transitions can occur *from* any ground state level except the $F = 2$, $M_F = 2$ level, because the $^2P_{1/2}$ has a maximum M_F of $+2$. On the other hand, emission from the excited state is governed by the selection rules

$$\begin{aligned} \Delta l &= -1 \\ \Delta M_F &= 0, 1 \\ \Delta F &= 0, 1 \qquad \Delta M = 0 \text{ and } \Delta F = 0 \text{ forbidden} \end{aligned} \tag{6-62}$$

These rules allow the $F = 2$, $M_F = 2$ level of the ground state to be fed by the emission process, but no absorption by the D_1 photons is allowed from that level. Hence, that level plays the role of our $+\frac{1}{2}$ state in the simple three-level example; in the absence of other processes, all atoms will end up in the (2, 2) level.

The most important process preventing complete polarization of the ground hfs states, aside from relaxation, is the interaction of D_2 photons with $+\hbar$ angular momentum absorbed from the (2, 2) ground state. The most successful operation of an optical pumping experiment requires an additional element between the lamp and the sample, a D_2 filter that passes D_1 but removes D_2 light. Such filters may be purchased commercially. Actually, the D_2 filter is not strictly necessary as long as a somewhat weaker ground state polarization is acceptable. The cell itself selectively filters D_2 light since the attenuation length for D_2 absorption is less than for D_1 absorption, both because there are twice as many ${}^2P_{3/2}$ states as there are ${}^2P_{1/2}$ states, and because the pumping action of the D_1 light makes the sample more transparent to D_1 light than to D_2 light. The number density of sample atoms may be chosen empirically (by varying the sample cell temperature) to achieve a maximum resonance signal. Under conditions of maximum signal, the magnetization, or polarization, within the sample is spatially inhomogeneous, and the description of the experiment requires the integral-differential Eq. (6-49) rather than the thin sample approximation, Eq. (6-50).

The magnetic resonance transitions are, of course, between the M_F sublevels of the ground state. In the strict zero field or Zeeman limit, the M_F levels of both $F = 1$ and $F = 2$ are equally spaced; the energy levels are linear functions of B_0. As $\mathbf{B}_0$ is increased, wavefunctions of different (F, M_F) are mixed into a particular state function since the coupling of the angular momentum vectors to $\mathbf{B}_0$ cannot be totally neglected compared to the coupling of **S** and **I**. The levels develop some curvature [except the (2, 2), (2, −2) and (1, 1) levels], and the spacing between adjacent levels is not constant. The single magnetic resonance line breaks up into $2I + 1$ lines, equally spaced. (We shall not prove it here. Quantum mechanics adepts can use perturbation theory to show it. Another way is to expand the Breit–Rabi formula [10], exact for all B_0, to keep only quadratic terms in B_0.) The $2I + 1$ lines may be resolved at remarkably low fields, on the order of 10 G, even with relatively primitive apparatus.

A variation of the pumping experiment that is possible in the more complex level structure of real atoms, as opposed to the three-level example, results in a population difference of the two $M_F = 0$ levels. This result is achieved with linearly polarized, or π pumping light. The selection rules for absorption are now restricted to $\Delta M_F = 0$ only, $\Delta M_F = 0$ and $\Delta F = 0$ forbidden. The emission selection rules are the same as before [Eq. 6-62]. The results are not intuitively evident, unless one draws with arrows all absorption and reemission transitions allowed on Fig. 6-14a. One sees that although the (2, 0) and (1, 0) levels have equal numbers of transitions *out*, the number of transitions *terminating* on the (2, 0) state is greater than the number terminating on the (1, 0) state. The equilibrium population of each of these states is, if we ignore the Boltzmann polarization, $(\frac{1}{8})N_0$, where N_0 is the total number of atoms in the sample cell. The ultimate populations after pumping, and in the absence of relaxation, are $0.162N_0$ for the (2, 0) and $1.137N_0$ for the (1, 0) levels. There is thus a substantial population difference between them, and the opacity of the cell can be increased in the usual way by causing rf transitions between the levels. We remark in passing that both the D_1 and D_2 lines contribute to the polarization of the $M_F = 0$ levels; no filters are needed.

The importance of achieving a population difference between the $M_F = 0$ levels lies in the relative field independence of the energy difference between these states. For example, the fractional shift in the frequency difference $\Delta\nu$ is approximately

$$\frac{\Delta\nu}{A} \sim 10^{-7}{B_0}^2$$

for ^{87}Rb, where $A = 6834.6826\cdots$ MHz and B_0 is in gauss. The other transitions are linearly proportional to B_0, with a coefficient on the order of 10^6 Hz/G. The line width of the transition is very narrow if the motional narrowing, discussed in Chapter 3, is fully effective. Hence, the other transitions are remote in frequency from the $0 \to 0$ transition even in $B_0 = 0.01$ G, so the dependence of the transition frequency on the external field may be as small as $1 : 10^4$.

The usefulness of such a device as a secondary frequency standard is obvious. The microwave oscillator producing the field that causes the (2, 0) to (1, 0) transition may be locked to the center of that resonance line by a feedback loop from the optical detector to a suitable frequency controlling point of the oscillator. (The oscillator is frequently a quartz crystal oscillator the frequency of which is "multiplied up" to the microwave range.) Short term stabilities better than $1 : 10^{12}$ are claimed for commercially available units using rubidium. The use of the term "secondary frequency standard" refers in part simply to the fact that the unit of time is not *defined* in terms of the period $1/A$, and in part to the fact that quite substantial frequency shifts are caused by a variety of experimental effects known as pressure shifts, light shifts, and wall shifts. The first and last of these refer to the change in the transition frequency caused by collisions of the alkali atom with buffer gas atoms and the walls of the container. The term "light shift" refers to a slight dependence of the oscillation frequency on the intensity of the pumping light. Such subtle effects become important when one has the capability of measuring to parts in 10^{12}. In fact, the hyperfine structure constant of hydrogen is the most accurately measured number in physics, having been measured with a random error of less than a tenth of a hertz out of 1420 MHz [11]. We have not used the hydrogen maser as the prototype of our discussion because the $(1, 0) \rightarrow (0, 0)$ population difference is achieved by an atomic beam technique rather than by optical pumping. Also, the detection of the resonance is done by monitoring directly the rf power.

One frequent application of optical pumping experiments is to the study of the ground state structure of other atoms whose excited state structures do not lend themselves directly to the optical pumping experiment. The first application of the technique that allows this to be done was to free electrons produced in an electrical discharge in the sample cell. The physical phenomenon that allows these experiments is the exchange interaction. In the collision between two atoms, or between an atom and an electron, there is a certain probability that the spins of each will be flipped during their interaction. The probability for this to happen is described in terms of an exchange cross section. Frequently, the exchange cross section is quite large, 10^{-13} to 10^{-14} cm^2, which is much greater than the geometrical cross section. If one of the atomic species involved in the collision is being optically pumped, then its net polarization is communicated to the other species, as one may readily see by using the principle of detailed balance. That is, the spin-flip collisions between the polarized (A) and unpolarized (B) atoms tend to polarize the latter (B) and unpolarize the former (A). But the A species is always having its polarization replenished by the action of the pumping light. Now, if the B atoms are unpolarized by a magnetic resonance experiment, they will depolarize the A atoms via the exchange collisions, rendering the sample more opaque to the pumping light. So the resonance of the spins may be detected through the two-link chain of A-B exchange and change in sample opacity. Such an experiment was first done by Dehmelt [12] with free electrons as B spins; it has since received wide application.

Optical pumping has been successful in polarizing systems other than the hyperfine structure levels of the ground electronic states. The most notable example is the case of the $^3S_1$ metastable level of helium [13]. That level is stable against decay to the $^1S_0$ ground state except by radiationless processes. A few $^3S_1$ metastables are present when an rf discharge is maintained in He gas at a few millimeters pressure. They are rather like distinct atoms in a buffer gas of He ground state atoms. In a stronger rf discharge, one of the spectral lines present with fair intensity is a triplet of wavelength about 1 μ from the $^3P_{2,1,0} \rightarrow {}^3S_1$ transition of the triplet series of helium. The energy-level diagram is shown in Fig. 6-15, with the effect of a small magnetic field also indicated.

The atoms being pumped are the He atoms in the 2 $^3S_1$ metastable state. They have a lifetime of about 10^{-4} sec in that level, and they decay to the

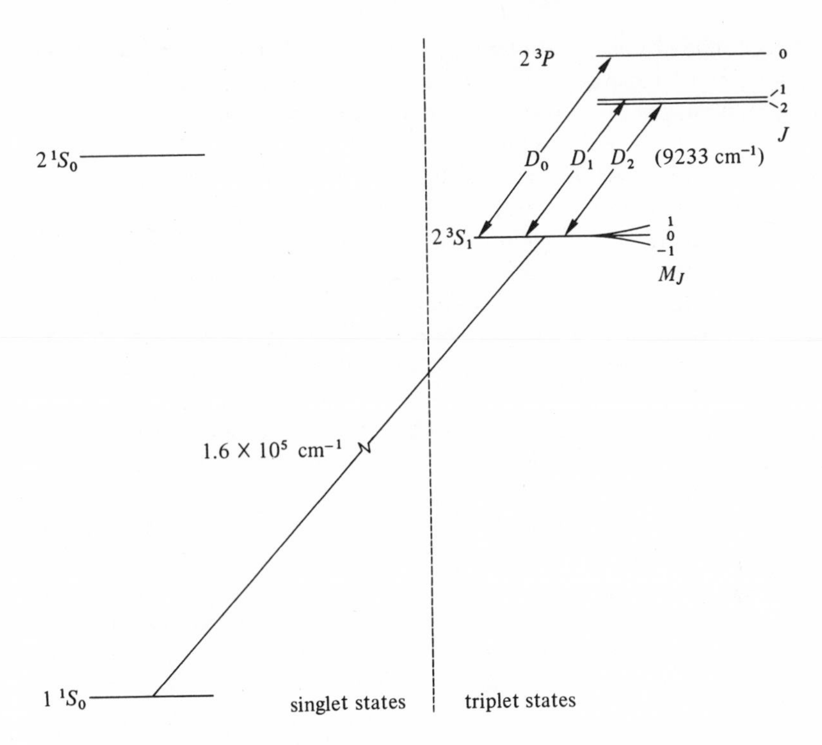

Fig. 6-15 The lower excited states of helium (not to scale). Only the Zeeman splitting of the $^3S_1$ state is shown.

true ground state of He, the $1\,^1S_0$ level, by a process depending somewhat on experimental conditions—collisions with the walls, for example. The pumping light consists of the three somewhat overlapping lines from the $2\,^3P$ triplet, marked D_0, D_1, and D_2 in Fig. 6-15. The polarization, or inequality of population, occurs among the three Zeeman levels of the $2\,^3S_1$ state. The experiment is otherwise much the same as in the alkali atoms, except that the details of the pumping process and the nature of the population imbalance achieved thereby are not as transparent as in the alkali case.

With unpolarized light, the $M_J = \pm 1$ levels of the $2\,^3S_1$ state are equally populated, but differ in population from the $M_J = 0$ level. Such a condition is usually termed *alignment*, the term *polarization* being reserved for situations in which a macroscopic magnetic moment is developed. Thus, in the experiments on alkali atoms described previously, the absorption of circularly polarized light *polarized* the Zeeman levels of the ground state, and the absorption of linearly polarized light produced an *alignment* of the ground state Zeeman levels. The same result occurs in the He case. In either event, rf transitions between the M_J Zeeman levels tend to produce equality of the populations of these levels, which is an effect in opposition to the effect of the pumping light, and hence it produces an increase in sample bulb attenuation of the pumping light, as usual.

We still have not seen how the alignment is produced. To do so, it is necessary to write down the probabilities for absorption to each of the 2 $^3P_2$ levels from each of the 2 $^3S_1$ levels. The relative absorption and spontaneous emission probabilities between any pair of magnetic sublevels is not the same, because absorption occurs only from unpolarized light propagating in the z direction, but emission can be in any direction with any polarization allowed by the selection rules. Thus, to give a particular example, the absorption probability for unpolarized light from the $^2S_1$, $M_J = 1$ level to the 2 $^3P_2$, $M_J = 1$ level is zero, but the spontaneous emission from that particular ^3P level back to the $M_J = 1$ ground level is $\frac{1}{2}$, the other $\frac{1}{2}$ going to the $M_J = 0$ level. An examination of the complete table of relative emission and absorption probabilities is necessary to be able to determine whether or not some alignment will occur. If the excited atom decays from the same fine structure level to which it was pumped, it is relatively hard to tell by simple inspection from such a table whether or not alignment will occur. At helium gas pressure of a few millimeters of mercury (i.e., a few Torr), the atom in the excited state undergoes enough collisions with the He buffer gas (the He atoms in the $^1S_0$ state) before it decays to scramble completely the ^3P level populations. All ^3P levels are equally populated in the decay, so that the rate at which the $^3S_1$ levels are fed can be read directly from the table, and it is more readily evident that alignment results.

6-3. MAGNETIC RESONANCE IN EXCITED STATES

It is quite evident that complete understanding of the atom or the nucleus must include the excited states of the system: their energies, their spins and parities, their magnetic properties—in effect, their wavefunctions. Magnetic resonance techniques on excited states have been employed almost from the very beginning of magnetic resonance to determine one or more of these properties of excited states except their gross energy splittings above the ground state. Optical, gamma-, and beta-ray spectroscopy serve well enough for that task. In addition, the use of excited states of nuclei as probes in solids has been an important tool in solid state physics, mainly through the Mössbauer effect. Because of the great variety of possibilities occasioned by the richness of possible questions, the lack of knowledge, and the technical feasibility of certain experiments, it is not conceivable that one can describe all the types of magnetic resonance experiments that have been done on excited states of atoms and nuclei in a few pages. We shall attempt to suggest some of the possibilities by discussing a model system having the twin advantages that it is the same as a real system on which a celebrated experiment has actually been done and that it has a clear classical analogy.

The system's ground state and relevant excited state are shown in Fig. 6-16. The ground state is a $^1S_0$ state, the excited state a $^3P_1$. The excited state decays to the ground state by electric dipole radiation with the emission of a left or right circularly polarized photon ($\sigma^{\pm}$) or a linearly polarized photon (π). (A real system that has this structure is mercury: the energy difference between the ground $6\,^1S_0$ and the excited $6\,^3P_1$ states gives rise to the 2537 Å resonance radiation line.) If a magnetic field is applied in the z direction, the excited state triplet is split, and, if the excited state lifetime is long enough, or the magnetic field strong enough, the energy differences can be resolved in the optical spectrum; that is, three lines are produced. This result is, of course, the Zeeman effect, and the virtue of this particular example is that the polarization properties of the spectral lines can be understood on the basis of the classical circulating charge picture of magnetic moments. In particular, the $\Delta M_J = \pm 1$ transitions give rise to circularly polarized light ($\sigma^{\pm}$ components) emitted in the $\pm z$ directions and linearly polarized light polarized in, say, the x direction for light emitted in the y direction. The π component is missing when one looks in the z direction, but it has z-direction polarization when emitted in the xy plane. All these rules for the angular distribution of the photons may be remembered by using the classically rotating (for σ) or linearly oscillating (for π) charge distribution as a sort of mnemonic device.

What are the questions one can ask about the system, and what answers can we extract from the energies, angular distribution, and polarization of the photons emitted in the decay? The nature of the answers depends on what we already know about the excited state (assume we know the ground state) and how we can prepare the excited state. Also, how complete is our measurement of the photon properties? Do we merely count photons,

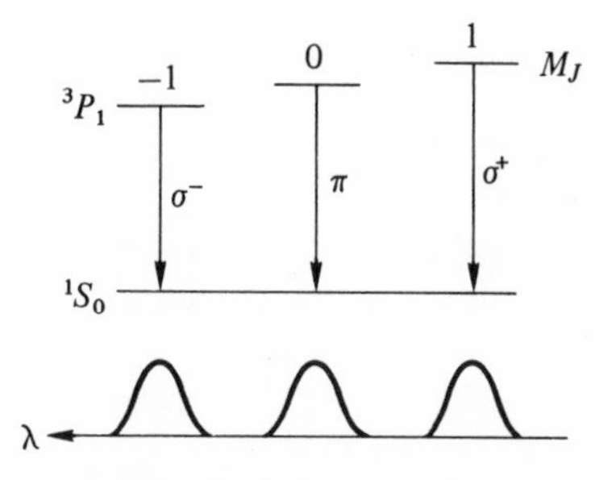

Fig. 6-16 Model system with $^3P_1$ excited state and $^1S_0$ ground state, with Zeeman effect and photon polarizations.

or do we measure their energy or their polarization; do we determine the complete angular distribution or some asymmetry in the angular distribution? Almost every conceivable combination of these possibilities has been used on some system or another—both atoms and nuclei. Let us look at some of the possibilities.

Suppose we count all photons emitted in the decay from the excited state without measuring their energy or polarization, and without knowing how the excited state was prepared. Clearly the measurement is a superposition of the decays from each excited state sublevel, with each level equally populated. The resultant distribution is isotropic in space, and we have learned nothing beyond the fact that the level had been excited and had decayed. Next, let us somehow arrange to excite only the $M_J = 0$ magnetic sublevel of the excited state in the presence of a field $\mathbf{B}_0$ in the z direction. We can know that we did that by measuring the angular distribution of the decay photon. We get the familiar dipole radiation pattern, with the number of counts proportional to $\sin^2\theta$, θ being measured from the z direction. We have learned that the decay is electric dipole from this measurement. Now let us take advantage of the angular anisotropy to measure the fine structure of the excited state—in this case, the Zeeman effect. The geometry of Fig. 6-17 will suffice. If the $M_J = 0$ level only is populated, the counter will not detect any photons. If an rf field at frequency

$$\hbar\omega_0 = g_J \beta_0 B_0 \tag{6-63}$$

is applied to the sample, transitions will be induced between the $M_J = 0$ and the $M_J = \pm 1$ states. The transition to the ground state can now occur via the $\Delta M_F = 1$ selection rule, and circularly polarized photons have a nonzero probability of being emitted in the $\theta = 0$ direction and of being counted. Thus, the excited state level structure is measured by the rf frequency, which causes the counting rate to increase in the z counter.

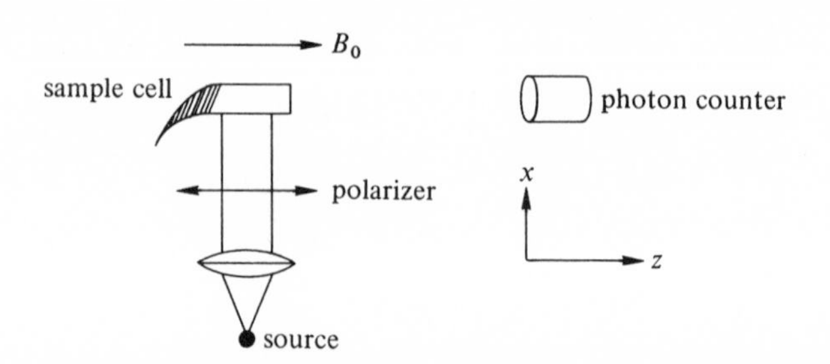

Fig. 6-17 Schematic experimental arrangement to detect magnetic resonance of 6 $^3P_1$ excited state of mercury.

We have described the essential features of the double resonance experiment of Brossel and Bitter on mercury [14]. The $M_J = 0$ level of the 6 $^3P_1$ state was populated by the method indicated in Fig. 6-17. Resonance radiation at 2537 Å from an Hg source was incident on the sample bulb and was polarized in the z direction. Thus, the absorption was of the $\Delta M_J = 0$ type, and the $M_J = 0$ state alone was excited. To add to the signal-to-noise in the actual experiment, π radiation in the y direction was also detected, and the *decrease* in that signal was correlated with the *increase* in the z direction signal at resonance. The curious hornlike sample cell is a traditional and necessary feature of many resonance radiation experiments. It is designed to prevent multiply reflected resonance radiation from reaching the detector.

Certain conditions have to be satisfied by the rf in any excited state resonance experiment. The rf field must be sufficiently strong at resonance to produce reasonable amplitudes of the $M_J = \pm 1$ states in the lifetime τ of the $^3P_1$ state. That is, we need to satisfy

$$\gamma H_1 \tau \simeq 1$$

In the mercury example, $\tau \sim 10^{-8}$ sec, and γ is the free electron value, 1.74×10^7 G^{-1} sec^{-1}; therefore, an H_1 of about 5 G is needed. Although that magnitude of H_1 is not difficult to achieve, it is already a substantial rf field. Were one working with a nuclear moment, the required H_1 would be enormous. We shall discuss that problem more later on.

The Brossel–Bitter technique of exciting one particular sublevel of the ^3P state works for this particular example. Let us think of the energy-level structure in more general terms again, and consider another alternative, one that has been feasible when the levels are the energy levels of nuclei. In Fig. 6-18, we have added a $^1S_0$ second excited state to the energy-level diagram of Fig. 6-16. The $^1S_0$ excited state decays to the $^3P_1$ state with emission of a $\sigma^{\pm}$ or π photon at ω, and the $^3P_1$ state decays as before. (Do not worry about all these singlet-triplet transitions, which

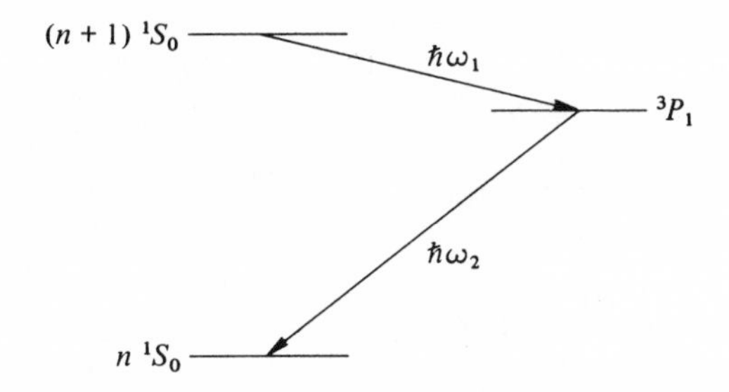

Fig. 6-18 Energy levels for a possible angular correlation experiment.

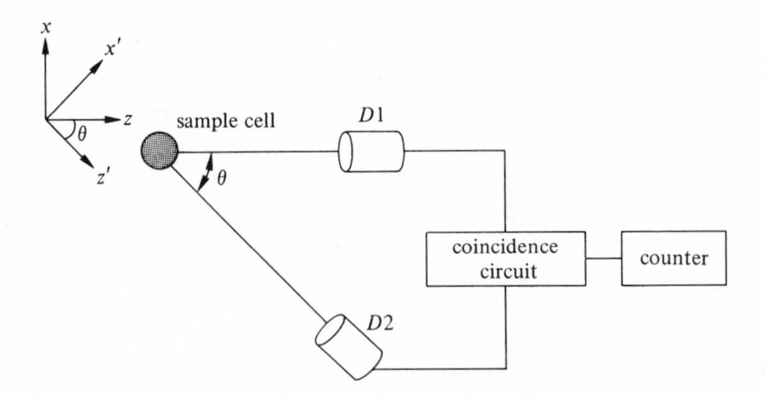

Fig. 6-19 Experimental arrangement for detecting angular correlations of successive photons emitted in cascade.

are supposedly forbidden under Russell–Saunders selection rules. For heavy atoms, *L-S* coupling is a poor approximation, and our description of the states does not include a considerable amount of configuration mixing.) Now consider the experimental arrangement of Fig. 6-19. Let the sample have atoms in various excited states (maintained by gaseous discharge, electron bombardment, or whatever). Let *D*1 detect only photons at ω_1, *D*2 only photons at ω_2 (although the frequency selectivity of *D*2 is probably unnecessary). Arrange the circuitry in the standard coincidence manner: the counter *D*2 registers a count only during a preselected time τ_2 (comparable to the lifetime τ of the ${}^3P_1$ level) after *D*1 has registered a photon of frequency ω_1. Thus, we are sure that *D*2 is detecting radiation from a $\Delta M_J = \pm 1$ transition from the ${}^3P_1$ level to the ${}^1S_0$ ground level. The reason we are sure of this is that *D*1 registers only $\sigma^{\pm}$ photons: note the direction of the field $\mathbf{B}_0$. So we have prepared the ${}^3P_1$ state in the $M_J = \pm 1$ states by the combination of geometry and the coincidence of *D*1 and *D*2. That is, *D*2 counts only those photons that come from the $M_J = \pm 1$ levels, since *D*2 is only activated by the decay at ω_1 that populates those levels.

If we measure the counting rate in *D*2 as a function of θ, call it $N_2(\theta)$, we can measure the angular distribution of $\sigma^{\pm}$ photons emitted in the ${}^3P_1 \rightarrow {}^1S_0$ decay. The angular distribution $N_2(\theta)$ can be worked out as follows (see Feynman [9], vol. 3, Chap. 18, for a detailed discussion of this argument). For a $\Delta M_J = \pm 1$ transition, the amplitude for emission of a photon in the $+z$ direction with $+\hbar$ intrinsic angular momentum is a; the amplitude for a $-\hbar$ photon emitted in the same direction in a $\Delta M_J = -1$ transition can be shown to be $-a$. Now our counter *D*2 defines a new direction, z', and it counts all photons incident upon it, independent of their polarization. So there are four ways or paths by which a photon can register in $D2$. We continue to describe the photon states in terms of angular momentum $\pm\hbar$ with respect to the direction of propagation, in this case the z' direction. A photon can arrive at $D2$ with $+\hbar$ in two ways, both via a $\Delta M_J' = 1$ transition, $M_J' = 1$ to $M_J' = 0$, where the prime indicates quantization with respect to the z' axis. One of these is from the amplitude or admixture of the $M_J' = +1$ state present in a wavefunction that is *entirely* $M_J = +1$ in the z-axis system. The other is via the admixture of $M_J' = +1$ present in a state that is *entirely* $M_J = -1$ in the z-axis system. Similar statements can be made for the $-\hbar$ photon. The admixture of $M_J' = +1$ present in the initial state of of $M_J = \pm 1$ is the matrix element

$$\langle + | R_y(\theta) | \pm \rangle = \tfrac{1}{2}(1 \pm \cos\theta) \tag{6-64}$$

where $R_y(\theta)$ is the rotation operator about the y axis. We collect the four amplitudes in a table in which the final photon state ($\pm\hbar$) is labeled by the rows, and the initial electron state in the z-axis system is labeled *over* the rows. Notice there is no decay involving the $M_J' = 0$ state, since that transition, $M_J' = 0$ to $M_J = 0$, is a π transition that always has zero amplitude in the direction of photon propagation, the z' direction.

In computing intensities—that is, the counting rate $N_2(\theta)$—there is always the difficult problem of whether to add amplitudes and take the absolute square of the sum, or vice versa. Table 6-1 is not the whole

Table 6-1

Amplitudes for Decay of Photon into Direction θ

	$M_J = 1$	$M_J = -1$
$+\hbar$	$\frac{a}{2}(1 + \cos\theta)$	$\frac{a}{2}(1 - \cos\theta)$
$-\hbar$	$-\frac{a}{2}(1 - \cos\theta)$	$-\frac{a}{2}(1 + \cos\theta)$

story of the process, since there is for each amplitude a time dependent phase factor. We should, in general, add amplitudes, square, and then average over the time dependent factors, since $D2$ does not register the time of our event but only that one has occurred. With random phases, the cross terms will vanish, and the final angular distribution will be the sum of the squares of the four entries to Table 6-1:

$$N_2(\theta) \propto |a|^2(1 + \cos^2\theta)[|c_1|^2 + |c_{-1}|^2] \tag{6-65}$$

where c_1 and c_{-1} are the amplitudes of the $M_J = \pm 1$ states, respectively, as determined by the initial decay.

We have described what is known as an angular correlation experiment. Let us recapitulate the various steps and emphasize their role. The intermediate state, $^3P_1$, is *prepared* by the act of counting the $^1S_0 \rightarrow {}^3P_1$ decay in the z direction. We prepared the state in either the $M_J = 1$ or -1 states so that $|c_1| = |c_{-1}|$, but did no experiment concerning the $M_J = 0$ level of the intermediate state. We could have prepared the initial state differently, by the setting of the coincidence circuitry with only $+\hbar$ photons on $D1$, for example [but we would have gotten the same $N_2(\theta)$]. In any event, the role of the first photon was to tell us something—not necessarily everything—about the intermediate state wavefunction. The angular dependence of the $D2$ counting rate was determined by the form of the rotation matrices of the intermediate state—that is, by how the intermediate state transforms under rotation. Hence, the angular correlation experiment has told us about the angular momentum properties of the intermediate state. Of course, we made use in an implicit way of the $^1S_0$ character of both the ground state and the initial excited state of the system. The decay $^1S_0 \rightarrow {}^3P_1$ prepared the intermediate state the way it did because the initial state was $^1S_0$, and the radiation was electric dipole. Similarly, the absence of a decay from the $M_J' = 0$ level of the intermediate state is determined by the second photon being electric dipole also. The important thing to realize is that even knowing that both the initial and ground states are $^1S_0$ is not enough to tell us what the intermediate state is—higher multipole decays are very common, particularly for nuclear excited states—but $N_2(\theta)$ tells us right away.

Now let us do a magnetic resonance experiment on the intermediate state. If we were to be complete, we would calculate the amplitudes $c_1(t)$, $c_0(t)$, and $c_{-1}(t)$ of the $M_J = 1$, 0, and -1 states in the presence of B_0 and the transverse rf field. The process is somewhat tedious, but clearly possible and rather elementary. The quantities $|c_1(t)|^2$ and $|c_{-1}(t)|^2$ in Eq. (6-65) will become smaller as the rf field induces transitions into the $M_J = 0$ state. The solution introduces a time dependence into the counting rate of $D2$ via the time dependence of c_1 and c_{-1}. We can guess the ultimate solution by recalling again that $D2$ does not, within broad limits, register *when* the photon arrives, only that it does. Hence, we would have to average $|c_1(t)|^2 + |c_{-1}(t)|^2$ over the time that $D2$ is accepting photons, say, a few intermediate state lifetimes. If the rf is strong enough to produce many radians of precession in the time that $D2$ accepts photons, we can assume that the time averaged amplitudes $c_1(t)$, $c_0(t)$, and $c_{-1}(t)$ are equal. Hence, with strong radio frequency,

$$|c_1(t)|^2 + |c_{-1}(t)|^2 = \tfrac{2}{3}$$

whereas the unperturbed sum was

$$|c_1(0)|^2 + |c_{-1}(0)|^2 = 1$$

Figure 6-20 shows a polar plot of $N_2(\theta)$ in the presence and absence of radio frequency. The angular correlation is not destroyed, just reduced in this example. The resonance is detected from the reduction in the counting rate for any angle. We should also point out that to calculate the line shape it is necessary to follow the complete program of finding $c_m(t)$ for arbitrary rf frequency and then averaging over time. It is frequently found that for strong rf field, the largest effect on the counting rate is produced at frequencies other than exact resonance, so that the

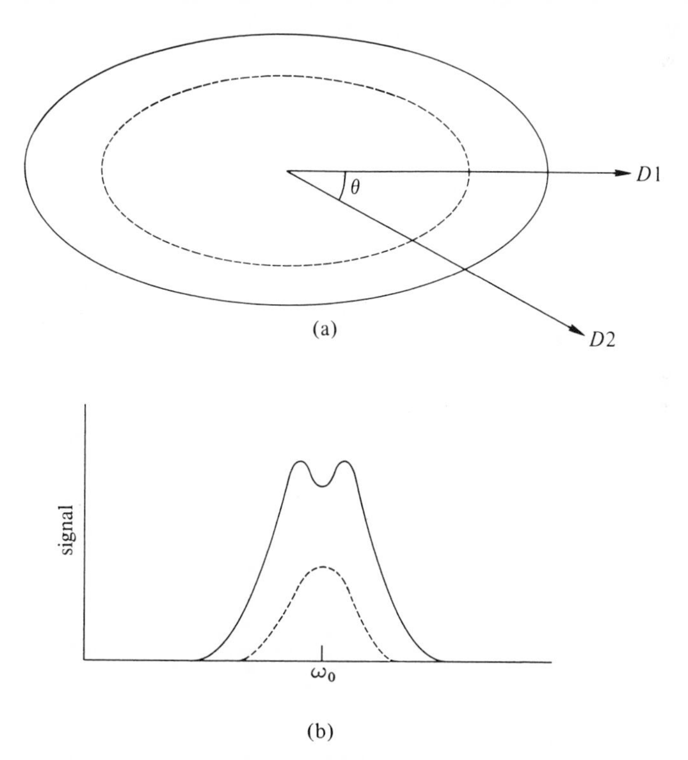

Fig. 6-20 (a) Angular correlation pattern of the text example with (solid) and without (dashed) rf induced transitions. (b) Resonance line shapes for electric dipole radiation. Solid: large rf field. Dashed: small rf field.

resonance curve is double peaked with a dip at exact resonance. More complex shapes are found for higher multipole radiation. In the other limit of relatively small rf field, a smaller change in $N_2(\theta)$ is, of course, produced, but the line shape becomes single peaked. The line shapes in the two limits are shown in Fig. 6-20b. In more complicated examples, the line shapes depend also on θ, and even for this system the detailed results depend on how the intermediate state is prepared—that is, on the angle the first photon propagation direction makes with $\mathbf{B}_0$.

We have presented this somewhat academic example in considerable detail because in the seemingly infinite variety of different situations arising in practice, the student can easily be lost unless he has available an understandable situation to which he returns for comfort and the analogies it may suggest. In the rest of this section, we shall concentrate on some examples from nuclear physics, in part to lend balance to the book but mostly because there seems to be rapidly increasing activity in such applications.

As we have emphasized, the applications are quite varied because almost all of them depend on some particular circumstance that allows an experiment to be done on an excited state. A general classification into types is given in Table 6-2 [15]. We have discussed the principle of the angular

Table 6-2

Methods Employing Radiative Detection of NMR

Method	Preparation	NMR sample	Detection
angular correlations	preceding radiation →	intermediate state: selected substates	change in angular correlation
nuclear orientation	H, low T or pumping →	oriented parent	change in counting rate of a single γ detector
nuclear reactions	incident beam →	oriented daughter	change in counting rate of a single γ detector
Mössbauer spectroscopy	polarized source →	selected substates	resonant absorption of γ ray by absorber

correlation experiment in some detail. We can best explain the rest of the table by considering briefly an example of each of the remaining methods. In fact, the field is so new that for some of the techniques only one or two examples exist, and for one of them it is not clear at this writing whether any successful experiments have yet been done.

The nuclear orientation technique is useful only for very long-lived excited states. The reason is that the technique relies on the attainment of thermal equilibrium between the spin of the excited state nucleus and the lattice of a solid at low temperature with, sometimes, an rf field producing the Overhauser effect or one of the similar dynamic orientation techniques we have not mentioned in this book. In the simplest, "brute force" method, the magnetic substates of the radioactive nucleus are populated unequally by applying a large enough magnetic field at low enough temperature to make the exponent, $\boldsymbol{\mu} \cdot \mathbf{H}/kT$ in the Boltzmann factor, on the order of unity. Practical difficulties prevent putting the nucleus in a solid sample and doing the cryogenics involved in less than a few minutes, so the method is limited to radioactive nuclei having lifetimes longer than this minimum time.

The recent applications of this method have been successful because the radioactive samples have been incorporated into ferromagnetic metals, usually iron. The reasons are to be found in the numbers needed to make $\boldsymbol{\mu} \cdot \mathbf{H}/kT < 1$. For nuclear moments, $\boldsymbol{\mu} \cdot \mathbf{H}$ is on the order of 3×10^{-4} cm^{-1} per 10,000 G. In those units, k, Boltzmann's constant, is about $(\frac{2}{3})$ $\text{cm}^{-1}/°\text{K}$. To achieve reasonable polarization, then, requires *both* millidegree temperatures (milli-Kelvins, in the newly defined terminology) and fields in the 10^5-G range. The low temperatures are achieved by the usual techniques of cryogenics—adiabatic demagnetization or helium dilution refrigerator. The implantation technique [16] or diffusion of the radioactive nucleus into a ferromagnet provides the high field, which has been found by nuclear magnetic resonance or Mössbauer effect experiments to be in the (1-5) $\times$ 10^5-G range at nuclei in ferromagnets. The internal fields at nuclei in ferromagnets must be determined for each impurity in each ferromagnet by inference from systematic studies. The fields are also not quite the same for each radioactive nucleus, so the resonance line is inhomogeneously broadened by 1% or so.

Figure 6-21 shows schematically an experimental arrangement and the energy-level diagram for a spin $I = 1$ nucleus in the internal field of the ferromagnet. Not shown is the very extensive cryogenic apparatus necessary to achieve temperatures in the milli-Kelvin range. The external field B_0 produced by the Helmholtz pair is on the order of 1 kG or so; it is applied in order to magnetically saturate the ferromagnetic foil, so that the direction of the field at each radioactive nucleus is the same. We can quickly analyze the experiment based on a hypothetical $I = 1$ nucleus

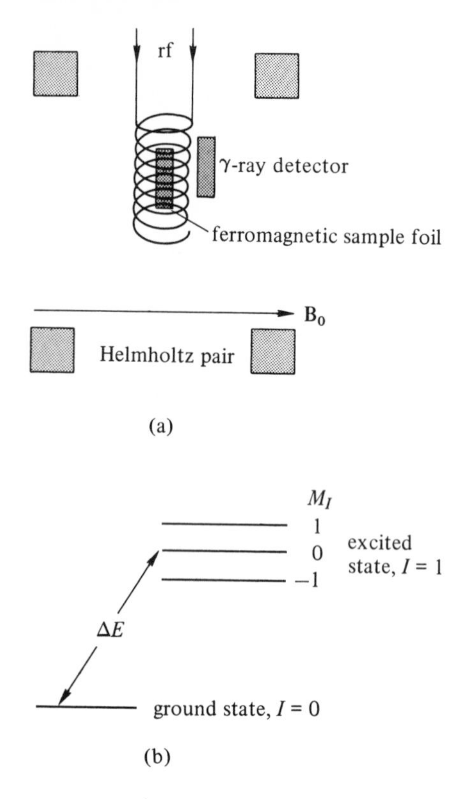

Fig. 6-21 (a) Experimental arrangement of fields, sample, and detector in a nuclear orientation NMR experiment. (b) Energy-level diagram, including Zeeman splitting (not to scale).

decaying to an $I = 0$ ground state with the emission of a several hundred-kilovolt γ ray. This example may be unrealistic in practice, since we have assumed the excited state to decay by electric dipole radiation, a particularly rapid decay mode, in contradiction to our requirements that the excited state live a long time. But for purposes of instructing with a simple and familiar system, we shall lay aside that *caveat* and proceed. The general expression for the counting rate at the detector, placed at $\theta = 0$ in Fig. 6-21, is

$$R = \sum_{m_I=-1}^{1} P_{m_I}(\theta) N(m_I) \tag{6-66}$$

where $P(\theta)$ is the probability that a photon is emitted from state m_I in direction θ, and $N(m_I)$ is the relative population of the magnetic substate m_I:

$$N(m_I) = \frac{\exp(-m_I \gamma \hbar H_0/kT)}{\sum_{m_I=-I}^{I} \exp(-m_I \gamma \hbar H_0/kT)} \tag{6-67}$$

In the present case, $P_0(0) = 0$, since π transitions in electric dipole radiation have zero amplitude for emission in the $\theta = 0$ direction. Notice that the counting rate depends on T, so that by itself the detector can measure the spin temperature T, or at least monitor changes in T. When the resonance condition is satisfied with a strong rf field, all levels become equally populated, and the counting rate in our example goes to

$$R' = \tfrac{1}{3}N_0[P_1(0) + P_{-1}(0)] \tag{6-68}$$

The observed signal is then $R' - R$. Note that if the rf field is removed, the counting rate R' will return to the original rate R with a time constant determined by the spin-lattice relaxation time. Spin-lattice relaxation times have indeed been measured by this method. The general theory of the resonance signal is, as usual, much more complicated, but no new ideas are involved. The signal from ^{125}Sb in Fe is shown in Fig. 6-22. Note the width of the line—on the order of 2% of the resonant frequency. It was necessary to frequency modulate the radio frequency by ± 400 kHz to cover the inhomogeneously broadened line. There should be some

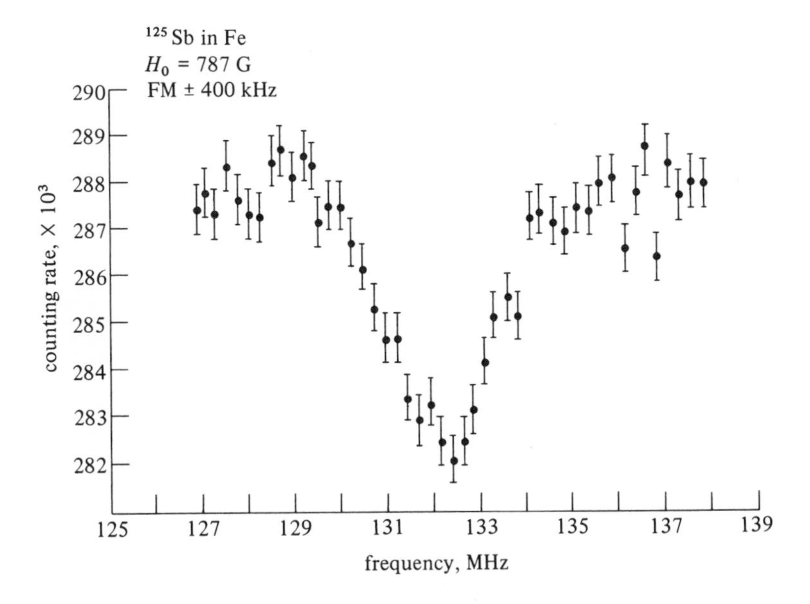

Fig. 6-22 Nuclear orientation NMR signal: ^{125}Sb in iron (after Matthias and Shirley [17]).

concern on the part of the reader that under these circumstances the magnitude of the radio frequency would not be great enough to equalize the level populations successfully, since it is applied for such a short time at the resonance frequency of each radioactive nucleus. There should also be some concern about the magnitude of the rf power dissipated in the sample, because a very small amount of heat input at low temperature can cause a substantial temperature rise. The use of a ferromagnetic host for the radioactive nuclei solves the problem of achieving a large rf field with small rf power by means of the phenomenon of "hyperfine enhancement." Figure 6-23 provides the basis for understanding this simple but important aspect of these experiments. The total instantaneous field at the nucleus is the vector sum of the small static field H_0, the applied rf field H_1, at right angles to H_0, and the hyperfine field H_{hfs} of several hundred kilogauss produced by the unpaired electron spins of the ferromagnetic sample. The relaxation time of the electron spins is short enough so that H_{hfs} can follow the instantaneous field $\mathbf{H}_0 + \mathbf{H}_1$ and is collinear with the resultant. Hence, the small component of $\mathbf{H}_{\text{total}}$ that is perpendicular to $\mathbf{H}_0$ is given approximately by

$$H_\perp = H_{\text{hfs}}\left(\frac{H_1}{H_0}\right) + H_1 \tag{6-69}$$

or, the effective H_1 is enhanced by the hyperfine enhancement factor F:

$$F = 1 + \frac{H_{\text{hfs}}}{H_0} \tag{6-70}$$

and F can easily be several hundred.

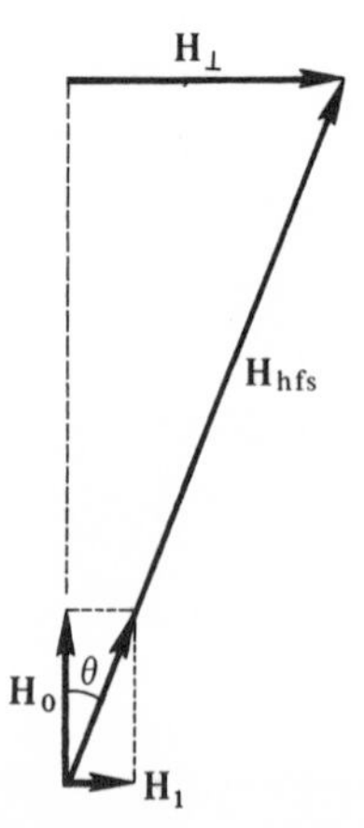

Fig. 6-23 Relations between the static applied field $\mathbf{H}_0$, the applied rf field $\mathbf{H}$, the hyperfine field $\mathbf{H}_{\text{hfs}}$, and the effective perpendicular field $\mathbf{H}_\perp$.

The large value of F has been used to advantage in an angular correlation experiment with a very short-lived intermediate state. The necessity of producing rf transitions between the magnetic sublevels of the intermediate state in the lifetime of that state requires $\gamma H_{\perp}\tau \simeq 1$. In the first reported experiment on angular correlation NMR [18] on the intermediate state of a nuclear γ-γ correlation measurement, the magnetic sublevel splitting of the intermediate state and the enhancement of the applied rf were accomplished by the large hyperfine field produced by a ferromagnet. The nucleus was ^{100}Rh, which begins its life as a radioactive source in this experiment as ^{100}Pd, an isotope with a 4-day half-life. Via electron capture, ^{100}Pd decays to ^{100m} Rh, an excited state of spin $I = 1$, 158.8 keV above the ground state. The decay to the ground state is via two successive γ rays. The first is to a spin $I = 2$ state with half-life of 235 nsec, the second to the spin $I = 1$ ground state. Thus, the situation is slightly more complicated than the angular momentum 0, 1, 0 sequence of our first Hg example, but electric dipole radiation is still involved. The counters were placed at $\theta = 0$ and 180° with respect to the field direction, so only σ components with angular distribution $(1 + \cos^2 \theta)$ were measured, just as in the Hg example. Resonances were observed corresponding to the partial destruction of the angular correlation both in Ni at 335 MHz and in Fe at 881 MHz. The rf enhancement factor F was about 10^3 in both cases. For a 235-nsec lifetime, and a nuclear magnetogyric ratio $|\gamma| \simeq 10^4$, the condition $\gamma H_1 \tau_{1/2} = 1$ requires an H_1 of 500 G, which, in turn, would require tens of kilowatts of rf power without the enhancement factor. With that factor, the experiment was done with a relatively modest 50 to 100 W. Perhaps the most spectacular number of the whole experiment was the estimate that 10^4 nuclei contributed to the observed signal. That number should be compared to the 10^{20} nuclei or so necessary to see a signal by the normal method of direct detection by bulk magnetization.

The third method listed in Table 6-2 represents a considerable variation of the theme of the first two. That it shares the same basic features is already indicated by the organizational principle of the table itself. The method or orientation of the radioactive nucleus and the method of detection of the depolarization by the rf field are both quite different from the methods used in nuclear orientation and angular correlations. We shall confine our discussion to a particular real experiment, since it is difficult as usual to generalize about an area of activity just being started.

Sugimoto *et al.* measured the magnetic moment of ^{17}F and, in subsequent similar experiments, the moments of ^{12}B and ^{12}N [19]. The radioactive nucleus ^{17}F is a β^+ emitter with a 66-sec half-life. Its magnetic moment is of particular interest because it is one of the pair of mirror nuclei ^{17}O – ^{17}F. Mirror nuclei differ from each other only in the replacement of a proton in one by a neutron. In particular, the stable

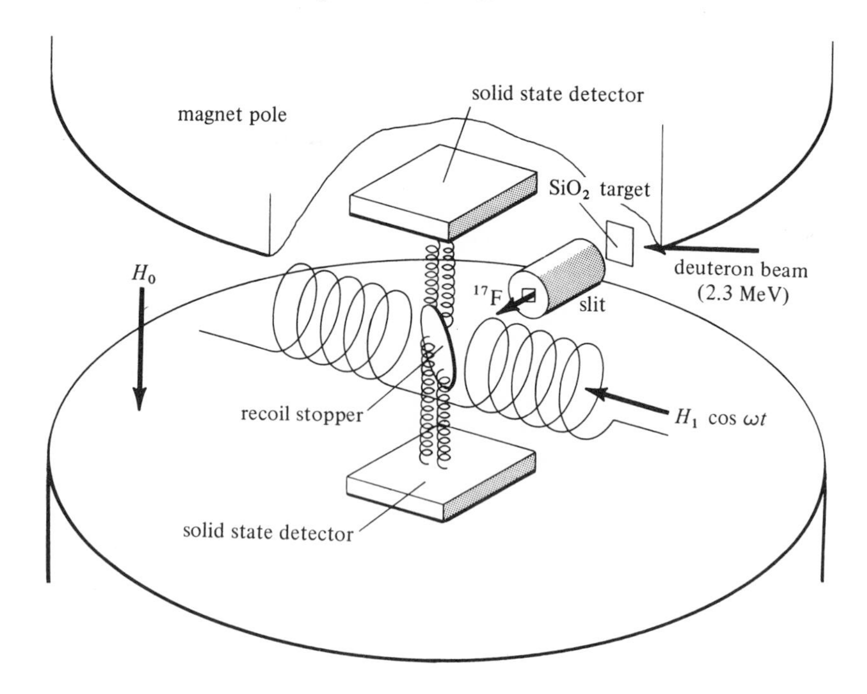

Fig. 6-24 A schematic drawing of the NMR experiment for ^{17}F.

nucleus ^{17}O has 8 protons and 9 neutrons and a spin $I = \frac{5}{2}$. The nucleus ^{17}F has 9 protons and 8 neutrons and a spin of $I = \frac{5}{2}$. The magnetic moments of the pair of mirror nuclei of mass 17 are of particular interest because they are formed by adding a neutron or a proton to the spin zero, even-even nucleus ^{16}O. The simplest theory of the magnetic moments of nuclei, the Schmidt model, ought to be particularly applicable to the nuclei of mass 15 and 17. For further discussion of the nuclear physics involved, the student should consult a suitable text on the subject, such as [20] or [21].

Nucleus ^{17}F in the experiment of Sugimoto *et al.* [19] is, in fact, formed by adding a proton to ^{16}O by means of the deuteron stripping reaction ^{16}O (d, n) ^{17}F.[4] That is, a proton has been *stripped* from the incident deuteron and added to the ^{16}O nucleus to form ^{17}F, while a neutron goes its way without the proton it came in with. The remarkable fact about the reaction is that the recoil nucleus, in this case ^{17}F, is polarized. The direction and extent of polarization depends on the direction the recoil nucleus goes relative to the incident beam. A schematic diagram of the experiment is indicated in Fig. 6-24. In that figure, the "reaction plane"

[4] The standard notation of nuclear scattering is to be read as follows: Target Nucleus A (incident particle, outgoing particle) Target Nucleus after reaction A'.

is defined by the incoming beam, which hits the SiO_2 target and the ^{17}F recoil momentum, selected by the collimating slit. The nuclear polarization is perpendicular to the reaction plane. The polarization is preserved by the magnetic field applied parallel to the polarization and, very surprisingly, is not entirely lost through the vicissitudes of the varying hyperfine fields, quadrupole fields, and atomic charge states to which it is subjected as it leaves the target. It is still preserved as the nuclei are stopped in the CaF_2 crystal that serves as the recoil stopper in Fig. 6-23. Two surprising features of the experiment so far cannot be explained adequately here: the existence of the polarization in the nuclear reaction and its maintenance afterward, even in the CaF_2 recoil stopper. The former phenomenon is obviously related to the conservation of angular momentum and the details of the d-^{16}O collision. The preservation of the polarization must be associated with the shortness of the time during which a particular depolarizing force is maintained. As we have seen, sufficiently short correlation times can be quite effective in reducing the effectiveness of large torques acting on nuclear moments.

The detection of the polarization of the recoil nuclei is accomplished by observing the asymmetry of the β decay. Because parity is not conserved in the β-decay process, the positron is not emitted isotropically but rather with an angular distribution given by

$$W(\theta) = 1 + \frac{\langle I_z \rangle}{I} \frac{v}{c} A_0 \cos \theta \qquad (6\text{-}71)$$

where θ is the angle between the direction of γ-ray emission and the polarization direction z, v is the velocity of the emitted β particle, and $(\langle I_z \rangle / I)$ is a measure of the degree of polarization of the ^{17}F nucleus. The coefficient A_0 depends on the details of the β decay, and is nearly unity for this particular example. The solid state detectors in Fig. 6-24 measure the counts in roughly the $\theta = 0$ and $\theta = 180°$ directions. The signal obtained by subtracting the counts from the two detectors is thus

$$R \simeq 2 \frac{\langle I_z \rangle}{I} \frac{v}{c} \qquad (6\text{-}72)$$

The experimenters found that the polarization of the recoil nuclei was maintained in the catcher foil for the β-decay half-life of ^{17}F. The SiO_2 target was bombarded for 20 sec; then, after a lapse of 10 sec to allow short-lived activity to die away, counts were registered for 90 sec while the rf field was turned on at a search frequency. The process was then repeated.

A similar experiment was done to measure the magnetic moment of ^{12}B. Experiments that use β-decay asymmetry to detect resonances have also been done with the polarization produced by a Stern–Gerlach magnet. Some of the experiments of this type by Commins and collaborators are particularly beautiful examples [22].

The Mössbauer effect measures with great precision the energy of the γ ray emitted from a radioactive nucleus, rather than the angular distribution. As we will show in an example, combined NMR and Mössbauer effect are distinct possibilities, and some experiments have been reported. However, at present writing, it is not yet clear whether the reported experiments have really been successful.

For our example, we shall consider the favorite iron nucleus ^{57}Fe in an electric field gradient. The excited state, $I = \frac{3}{2}$, decays with a half-life of 0.1 μsec to the spin $\frac{1}{2}$ ground state with the emission of a 14.4-keV γ ray. The excited state has a quadrupole moment, so the energy-level diagram with the high field quantum numbers m_I of the levels in an electric field gradient is as shown in Fig. 6-25. The two-line pattern of Fig. 6-25b originates, as seen in Fig. 6-25a, from the quadrupole splitting of the excited state. If a strong rf field were applied at the quadrupole frequency with sufficient strength so that $\gamma H_1 \tau_{1/2} \gg 1$, then each nucleus would no longer be in either the $m_I = \pm\frac{3}{2}$ or $m_I = \pm\frac{1}{2}$ state exclusively during the γ-ray decay time, and the energy of the γ ray would be the average of the two energies. Put somewhat better, the 14.4-keV γ ray would be frequency modulated at frequency γH_1 with a frequency deviation $\Delta E/\hbar$, so the modulation index is $\Delta E/\gamma H_1 \hbar$. The doublet will collapse into a single line if $\gamma H_1 \hbar \gg \Delta E$, an even more stringent condition than $\gamma H_1 \tau_{1/2} \gg 1$. But these estimates are fantasy, since even $\gamma H_1 \tau_{1/2} \simeq 1$ requires a kilogauss rf field, which is impractical.

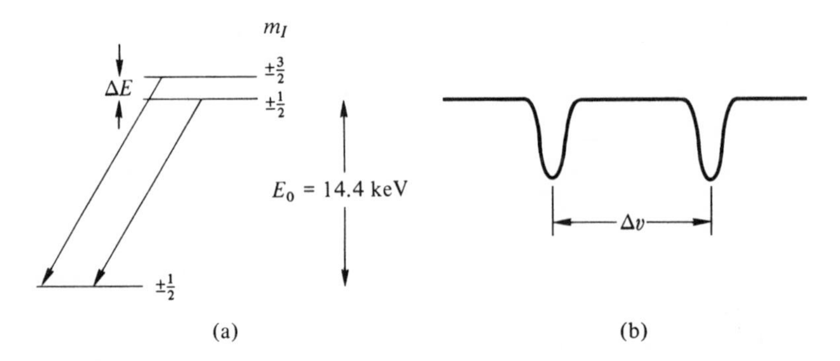

Fig. 6-25 (a) Energy-level diagram of ^{57}Fe ground and 14.4 keV excited states, showing splitting if $I = \frac{3}{2}$ excited state in an electric field gradient. (b) Mössbauer spectrum. The doublet is split by the velocity $\Delta v = c\,\Delta E/E_0$.

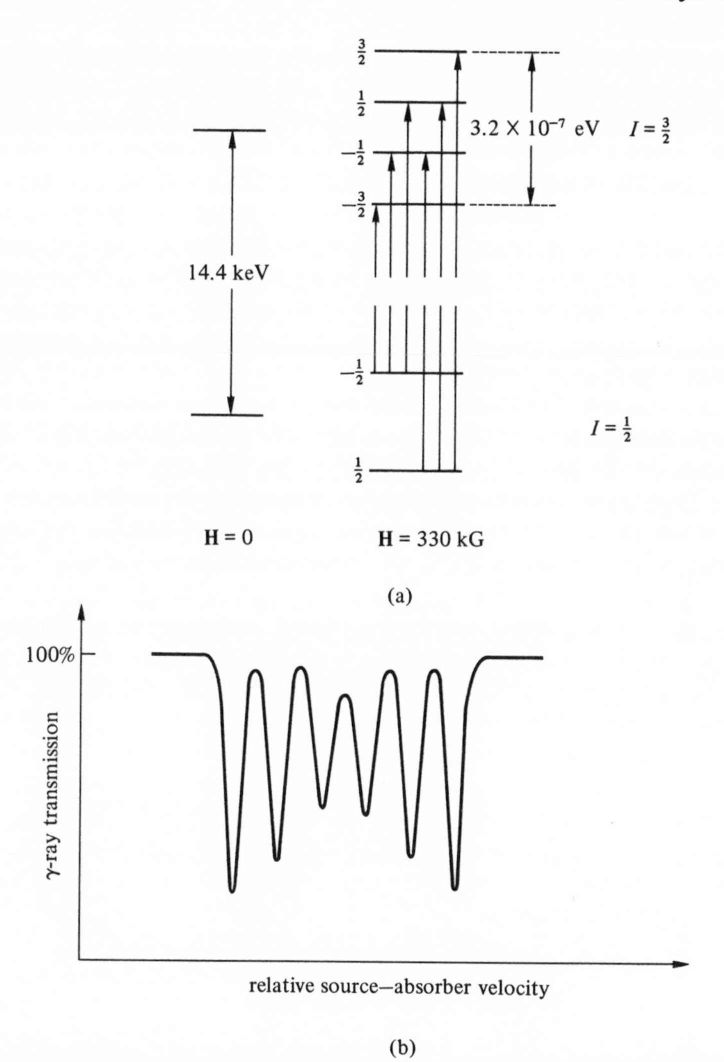

Fig. 6-26 (a) ^{57}Fe energy-level diagram for a large internal magnetic field of a ferromagnet, including allowed transitions. (b) Mössbauer spectrum with iron as the source, and a single line absorber.

The first tries at observing the effects have made use of the hyperfine enhancement of the rf field by using samples of ferromagnetic iron. The Mössbauer pattern is in this case a characteristic six-line pattern, with both the ground state and excited state split by the 500-kG hyperfine field. Nuclear resonance is possible at either the ground state or excited state frequencies (Fig. 6-26). Application of relatively modest rf fields has been observed to produce a change in the Mössbauer pattern, but there is evidence that the effect is largely caused by a magnetostrictive modulation of the sample by the rf field, and the observed effect is from velocity modulation of the sample rather than from nuclear transitions [23].

Somewhat different possibilities exist in the case of isolated Fe^{3+} ions in paramagnetic salts. If the electron spin relaxation time is long enough, three overlapping patterns of the type of Fig. 6-26 are seen. The strong field is produced by the hyperfine interaction between the ^{57}Fe nucleus and the five 3*d* electrons of the Fe^{3+} ion, which combine to form a total spin of $\frac{5}{2}$. If the electronic spin stays constant in space long enough—if it preserves itself long enough in one of the three Kramers degenerate states $\pm\frac{1}{2}$, $\pm\frac{3}{2}$, $\pm\frac{5}{2}$ in zero applied field—the nuclei see hyperfine fields with magnitudes in the ratio 1:3:5. Rapid relaxation of the electronic spin washes out the pattern seen for nearly static spins. Rapid transitions among the electronic states could also be caused by an rf field acting on the electronic spins. The large electronic moment would put the required H_1 within reason. Used this way, the Mössbauer effect would simply be a paramagnetic resonance detector. It is not clear that the experiment would have any advantages over the standard paramagnetic resonance detection, but it is unwise to dismiss in advance experiments that are receiving serious consideration. One should remember that Otto Stern never found Rabi's resonance experiments to his taste, but they did have some interesting consequences, after all.

6-4. LITERATURE GUIDE

The selection of introductory and intermediate level literature on the material of this chapter is quite thin except for optical pumping. The introductory notes and reprint collection of Bernheim [8] has already been mentioned. "The Resource Letter on Lasers and Optical Pumping" [24] provides a fairly complete list of references prior to 1964. An article not listed there, by deZafra [25], has proven to be particularly readable for students.

One type of experiment having elements of both optical pumping and angular correlation experiments is the *level crossing* experiment [26]. We have not discussed it at all because it is not a resonance experiment—except, perhaps, at zero frequency—but it is becoming a frequently used adjunct to resonance experiments. It is discussed by Bernheim [8].

The only reference at this writing to the details of ion-cyclotron resonance experiments is the article in *Science* by Baldeschwieler [2]. The older applications of cyclotron resonance have been described in some elementary articles, such as the one in *Scientific American* [27], and in more comprehensive monographs, particularly in the *Solid State Physics* series [28]. These articles emphasize the solid state physics to be learned from the technique and require a considerable background in that subject. Similar statements hold for the application to metals by the Azbel'–Kaner method. Brief treatments similar to the one given in this chapter may be found in many introductory solid state physics texts [4], [5].

Introductory discussions of NMR in excited states, particularly nuclear excited states, are even more rare, since the techniques are new in practice, if not in concept. High level summaries of recent work and references to the original journal articles may be found in the *Proceedings of the Conference on Hyperfine Interactions*, held at Asilomar in August, 1967 [15]. Note particularly the articles in Section 10 of that volume.

Problems

6-1. Show that in the absence of the electric field, Eq. (6-10) may be written in the Bloch equation form,

$$\frac{d\mathbf{P}}{dt} = \gamma \mathbf{P} \times \mathbf{B}_0 - \frac{\mathbf{P}}{\tau_r}$$

where $\mathbf{P} = n\langle \mathbf{p} \rangle = nq\langle \mathbf{r} \rangle$ is the macroscopic electric dipole moment per unit volume, the electric polarization vector, and $\gamma = q/mc$. Notice that Eq. (6-10) cannot be cast into a form completely analogous to the Bloch equations because the electric dipole moment is not "rigid"; the electric field changes the magnitude as well as the direction of the quantity $\mathbf{p} = q\mathbf{r}$. That is, there is no Wigner–Eckart theorem for these dipoles.

6-2. Verify that the Larmor frequency of ^{85}Rb is 700 kHz/G.

6-3. Using the Briet–Rabi formula, verify the assertion on p. 185 that $\Delta\nu/A \sim 10^{-7}B_0^2$ for the field dependence of the frequency difference between the $F = 2$, $M_F = 0$ and $F = 1$, $M_F = 0$ hyperfine levels (B_0 is expressed in gauss).

6-4. Suppose the triplet of Zeeman split energy levels of Fig. 6-18 are not equally spaced. Devise the optimum angular correlation magnetic resonance experiment for measuring the different energies corresponding to the $M_J = 1$ to $M_J = 0$ energy difference and the $M_J = -1$ to $M_J = 0$ energy difference. Be sure the technique allows you to determine which splitting is larger.

6-5. Determine the effective mass of conduction electrons in potassium from Fig. 6-10b.

6-6. Find the Fermi momentum k_F from Fig. 6-11.

References

1. H. Sommer, H. A. Thomas, and J. A. Hipple, *Phys. Rev.* **82**, 697 (1951).
2. J. D. Baldeschwieler, *Science* **159**, 263 (1968).
3. W. Shockley, *Electrons and Holes in Semiconductors*, D. Van Nostrand Co., Inc., Princeton, New Jersey (1950).
4. C. Kittel, *Introduction to Solid State Physics*, 3rd ed., John Wiley & Sons, Inc., New York (1966).
5. J. M. Ziman, *Principles of the Theory of Solids*, Cambridge University Press, Cambridge, England (1964).
6. M. Ya. Azbel' and K. A. Kaner, *J. Phys. Chem. Solids* **6**, 113 (1958).
7. J. M. Ziman, *Electrons in Metals, A Short Guide to the Fermi Surface*, Taylor and Francis, Ltd., London (1963).
8. R. Bernheim, *Optical Pumping, An Introduction*, W. A. Benjamin, Inc., New York (1965).
9. R. P. Feynman, R. B. Leighton, and Matthew Sands, *The Feynman Lectures on Physics*, Addison-Wesley Publishing Co., Reading, Massachusetts, vol. 3 (1965).
10. G. Breit and I. I. Rabi, *Phys. Rev.* **38**, 2002 (1931).
11. S. B. Crampton, Daniel Kleppner, and Norman F. Ramsey, *Phys. Rev. Letters* **11** 338, (1963).
12. H. G. Dehmelt, *Phys. Rev.* **109**, 381 (1958).
13. P. A. Franken and F. D. Colegrove, *Phys. Rev. Letters* **1**, 316 (1958).
14. J. Brossel and F. Bitter, *Phys. Rev.* **86**, 308 (1952).
15. *Hyperfine Structure and Nuclear Radiations*, E. Matthias and D. A. Shirley, Eds., North Holland Publishing Company, Amsterdam (1968). See the article by D. A. Shirley, p. 843.
16. Reference 15, Section 8.
17. Reference 15, article by J. A. Barclay, W. D. Brewer, E. Matthias, and D. A. Shirley, p. 902.
18. Reference 15, article by E. Matthias, D. A. Shirley, N. Edelstein, H. J. Korner, and B. A. Olsen, p. 878.
19. Reference 15, article by K. Sugimoto, p. 859.
20. H. Enge, *Introduction to Nuclear Physics*, Addison-Wesley Publishing Co., Reading, Massachusetts (1966).
21. D. Halliday, *Introductory Nuclear Physics*, 2nd ed., John Wiley & Sons, Inc., New York (1955).
22. E. D. Commins and D. A. Dobson, *Phys. Rev. Letters* **10**, 347 (1963); E. D. Commins and H. R. Feldman, *Phys. Rev.* **131**, 700 (1963); F. P. Calaprice, E. D. Commins, and D. A. Dobson, *Phys. Rev.* **137**, B1453 (1965).
23. N. D. Heiman, L. Pfeiffer, and J. C. Walker, *Phys. Rev. Letters* **21**, 93 (1968).
24. H. W. Moos, *Am. J. Phys.* **32**, 589 (1964).
25. R. L. deZafra, *Am. J. Phys.* **28**, 646 (1960).
26. P. A. Franken, *Phys. Rev.* **121**, 508 (1961).
27. A. R. Mackintosh, *Sci. Am.* **110** (July 1963).
28. B. Lax and J. G. Mavroides, *Solid State Physics*, F. Seitz and D. Turnbull, Eds., Academic Press Inc., New York, vol. 11 (1961).

APPENDIX

Some Quantum Mechanics of Spin $\frac{1}{2}$

In a few crucial places in this book, it is important to be able to understand the quantum mechanics of a spin $\frac{1}{2}$ particle in magnetic fields. In this Appendix, we provide a resumé with emphasis on the mathematics of the description. A full appreciation of the physics of the subject requires study of more than a brief Appendix, and for this further study volume 3 of the Feynman *Lectures on Physics* is recommended.

Although it is clumsy to do so for a general angular momentum J, for spin $\frac{1}{2}$ it is most convenient to use the matrix notation, particularly the Pauli spin matrices. In that notation, the angular momentum operators are 2×2 matrices, and the spin wavefunctions and their complex conjugates are column and row vectors. Thus, a spin up is described by $\begin{pmatrix}1\\0\end{pmatrix}$, a spin down by $\begin{pmatrix}0\\1\end{pmatrix}$. In purely mathematical language, these vectors are the basis vectors that we use to span the vector space. A vector in this space is written in general as

$$\alpha\begin{pmatrix}1\\0\end{pmatrix} + \beta\begin{pmatrix}0\\1\end{pmatrix} = \begin{pmatrix}\alpha\\\beta\end{pmatrix} \tag{A-1}$$

where α and β may be complex. The vector $\begin{pmatrix}\alpha\\\beta\end{pmatrix}$ is often conveniently written in the " bra " and " ket " notation of Dirac as

$$|\chi\rangle = \begin{pmatrix}\alpha\\\beta\end{pmatrix} \quad \text{(" ket " vector)} \tag{A-2}$$

The operation of forming the scalar product in this space is performed by defining the " bra " vector

$$\langle\chi| = (\alpha^*\beta^*) \tag{A-3}$$

as a row vector in the matrix notation. Thus, the scalar product of a vector with itself, its length, is given by

$$\langle \chi | \chi \rangle = (\alpha^* \beta^*) \begin{pmatrix} \alpha \\ \beta \end{pmatrix} = \alpha^* \alpha + \beta^* \beta \tag{A-4}$$

The scalar product of two different vectors

$$|\chi_1\rangle = \begin{pmatrix} \alpha_1 \\ \beta_1 \end{pmatrix} \qquad |\chi_2\rangle = \begin{pmatrix} \alpha_2 \\ \beta_2 \end{pmatrix}$$

is given by

$$\langle \chi_1 | \chi_2 \rangle = (\alpha_1^* \beta_1^*) \begin{pmatrix} \alpha_2 \\ \beta_2 \end{pmatrix} = \alpha_1^* \alpha_2 + \beta_1^* \beta_2 \tag{A-5}$$

The order matters:

$$\langle \chi_2 | \chi_1 \rangle = \alpha_2^* \alpha_1 + \beta_2^* \beta_2^* = \langle \chi_1 | \chi_2 \rangle^*$$

The orthogonality of the basis vectors is obvious in this notation:

$$\langle + | - \rangle = (10) \begin{pmatrix} 0 \\ 1 \end{pmatrix} = \langle - | + \rangle = 0$$

Operator J_z in this notation is the matrix

$$J_z = \frac{\hbar}{2} \begin{pmatrix} 1 & 0 \\ 0 & -1 \end{pmatrix} = \frac{\hbar}{2} \sigma_z \tag{A-6}$$

where the z component of the Pauli spin matrix has been defined. The basis states are eigenstates of J_z:

$$\frac{\hbar}{2} \begin{pmatrix} 1 & 0 \\ 0 & -1 \end{pmatrix} \begin{pmatrix} 1 \\ 0 \end{pmatrix} = \frac{\hbar}{2} \begin{pmatrix} 1 \\ 0 \end{pmatrix}$$

and

$$\frac{\hbar}{2} \begin{pmatrix} 1 & 0 \\ 0 & -1 \end{pmatrix} \begin{pmatrix} 0 \\ 1 \end{pmatrix} = -\frac{\hbar}{2} \begin{pmatrix} 0 \\ 1 \end{pmatrix} \tag{A-7}$$

with eigenvalues $\pm\hbar/2$, respectively.

Before exploring the other components of angular momentum, let us use the preceding notation to solve the problem of a spin in a magnetic field in the z direction. The Hamiltonian is

$$\mathscr{H} = -\gamma\hbar J_z H_0 = -\frac{\gamma\hbar H_0}{2}\begin{pmatrix}1 & 0\\ 0 & -1\end{pmatrix} \tag{A-8}$$

Schrödinger's equation is

$$i\hbar\begin{pmatrix}\dot{\alpha}(t)\\ \dot{\beta}(t)\end{pmatrix} = -\frac{\gamma\hbar H_0}{2}\begin{pmatrix}1 & 0\\ 0 & -1\end{pmatrix}\begin{pmatrix}\alpha(t)\\ \beta(t)\end{pmatrix} \tag{A-9}$$

Equation (A-9) reduces to two independent, first-order, differential equations, one for each component of the vector:

$$\begin{aligned} i\hbar\dot{\alpha} &= -\frac{\gamma\hbar H_0}{2}\,\alpha(t) \\ i\hbar\dot{\beta} &= +\frac{\gamma\hbar H_0}{2}\,\beta(t) \end{aligned} \tag{A-10}$$

The solutions are readily found:

$$\begin{aligned} \alpha(t) &= \alpha(0)\exp\left(\frac{i\gamma H_0}{2}\,t\right) \\ \beta(t) &= \beta(0)\exp\left(\frac{-i\gamma H_0}{2}\,t\right) \end{aligned} \tag{A-11}$$

Equations (A-11) express, of course, the well-known result that the energy difference between the spin-up and spin-down states is $\hbar\gamma H_0$.

It is equally important to us to see in what way Eqs. (A-11) describe the classical precession of a spin about the field direction. To do so, we must ask how to find the eigenstates of the operators $J_x = (\hbar/2)\sigma_x$ and $J_y = (\hbar/2)\sigma_y$. The Pauli spin matrices σ_x and σ_y are given by

$$\sigma_x = \begin{pmatrix}0 & 1\\ 1 & 0\end{pmatrix} \qquad \sigma_y = \begin{pmatrix}0 & -i\\ i & 0\end{pmatrix} \tag{A-12}$$

For this simplest of all cases, spin $\frac{1}{2}$, we can find the answer by analogy.

The properties of the basis vectors $|+\rangle = \begin{pmatrix}1\\0\end{pmatrix}$ and $|-\rangle = \begin{pmatrix}0\\1\end{pmatrix}$ that we must formally recognize are Eqs. (A-7), expressible in brief as

$$\sigma_z\,|\pm\rangle = \pm|\pm\rangle \tag{A-13}$$

and that the probability of finding a spin in the up (down) state is unity for the $|+\rangle$ ($|-\rangle$) eigenstates:

$$|\langle\pm\,|\,\pm\rangle|^2 = 1 \tag{A-14}$$

Our problem is to find the correct linear combinations of the $|+\rangle$ and $|-\rangle$ states having the same property with respect to σ_x and σ_y:

$$\sigma_x\,|x\rangle = |x\rangle \qquad \sigma_y\,|y\rangle = |y\rangle \tag{A-15}$$

That is, we want to find α_1, β_1, α_2, and β_2, such that

$$|x\rangle = \begin{pmatrix}\alpha_1\\\beta_1\end{pmatrix} \qquad |y\rangle = \begin{pmatrix}\alpha_2\\\beta_2\end{pmatrix}$$

and property (A-15) is satisfied. In matrix form, we have

$$\begin{pmatrix}0 & 1\\1 & 0\end{pmatrix}\begin{pmatrix}\alpha_1\\\beta_1\end{pmatrix} = \begin{pmatrix}\beta_1\\\alpha_1\end{pmatrix}$$

or

$$\begin{pmatrix}\beta_1\\\alpha_1\end{pmatrix} = \begin{pmatrix}\alpha_1\\\beta_1\end{pmatrix}$$

which implies $|\alpha_1| = |\beta_1| = 1/\sqrt{2}$, and

$$|x\rangle = 2^{-1/2}e^{i\delta_x}\begin{pmatrix}1\\1\end{pmatrix} \tag{A-16}$$

We need not specify the phase δ_x. Similarly, the requirement of the y equations yields

$$|y\rangle = 2^{-1/2}e^{i\delta_y}\begin{pmatrix}1\\i\end{pmatrix} \tag{A-17}$$

Given these wavefunctions, we can calculate the probability that the spin described by Eqs. (A-11) points in the x direction by calculating

$$|\langle x\,|\,\chi(t)\rangle|^2 = \left|\frac{e^{-i\delta_x}}{\sqrt{2}}(1 \quad 1)\begin{pmatrix}\alpha(0)\exp(i\omega_0 t/2)\\ \beta(0)\exp(-i\omega_0 t/2)\end{pmatrix}\right|^2$$

$$= \frac{1}{2}\left|\alpha(0)\exp\frac{i\omega_0 t}{2} + \beta(0)\exp\frac{-i\omega_0 t}{2}\right|^2$$

$$= \frac{1}{2}[|\alpha(0)|^2 + |\beta(0)|^2 + \beta\alpha^*\exp(-i\omega_0 t) + \alpha\beta^*\exp(i\omega_0 t)] \tag{A-18}$$

We shall be able to unscramble the meaning of Eq. (A-18) by choosing for an example the initial conditions $\alpha(0) = \beta(0) = 1/\sqrt{2}$. Then,

$$|\langle x\,|\,\chi(t)\rangle|^2 = \frac{1 + \cos\omega_0 t}{2} \tag{A-19}$$

For

$$|\langle y\,|\,\chi(t)\rangle|^2,$$

we find

$$|\langle y\,|\chi(t)\rangle|^2 = \left|\frac{e^{-i\delta_y}}{\sqrt{2}}(1 \quad i)\begin{pmatrix}\alpha(0)\exp(i\omega_0 t/2)\\ \beta(0)\exp(-i\omega_0 t/2)\end{pmatrix}\right|^2$$

$$= \frac{1}{2}\left|\alpha(0)\exp\frac{i\omega_0 t}{2} + i\beta(0)\exp\frac{-i\omega_0 t}{2}\right|^2$$

$$= \frac{1}{2}(1 - \sin\omega_0 t) \tag{A-20}$$

where for the last equality we have again chosen $\alpha(0) = \beta(0) = 1/\sqrt{2}$. This particular choice of initial conditions allows us to make the immediate identification with the classically precessing spin, precessing about the z axis with angular frequency $\omega_0 = \gamma H_0$, since we note that the probability of finding the spin in the x direction is unity at times such that $\omega_0 t = 0$, 2π, 4π, ..., and in the y direction is unity at the *later* times $\omega_0 t = 3\pi/2$, $7\pi/2$, $11\pi/2$, ...: three-quarters of a period later. (The spin precesses from x to $-y$ to $-x$ to y to x; see Fig. 1-4.)

As a final exercise in the quantum mechanics of the spin $\frac{1}{2}$, we solve the magnetic resonance problem of a free spin starting in the spin-up state in the presence of a static field H_0 in the z direction and a transverse rotating field

$$\mathbf{H}_1 = H_1(\hat{\mathbf{i}} \cos \omega t - \hat{\mathbf{j}} \sin \omega t)$$

Schrödinger's equation is

$$-\frac{\gamma\hbar}{2}\boldsymbol{\sigma}\cdot\mathbf{H}|\chi\rangle = i\hbar\,|\dot{\chi}\rangle \tag{A-21}$$

In matrix form,

$$-\frac{\hbar}{2}\begin{pmatrix}\omega_0 & \omega_1 e^{i\omega t}\\ \omega_1 e^{-i\omega t} & -\omega_0\end{pmatrix}\begin{pmatrix}\alpha(t)\\ \beta(t)\end{pmatrix} = i\hbar\begin{pmatrix}\dot{\alpha}\\ \dot{\beta}\end{pmatrix} \tag{A-22}$$

where $\omega_1 = \gamma H_1$, $\omega_0 = \gamma H_0$.

The two differential equations one obtains are

$$-\frac{\omega_0}{2}\alpha - \frac{\omega_1}{2}e^{i\omega t}\beta = i\dot{\alpha} \tag{A-23a}$$

$$-\frac{\omega_1}{2}e^{-i\omega t}\alpha + \frac{\omega_0}{2} = i\dot{\beta} \tag{A-23b}$$

The initial condition is $|\chi(0)\rangle = \begin{pmatrix}1\\0\end{pmatrix}$. One wishes to find the transition probability to the down state, or $|\beta(t)|^2$. The most direct procedure is to eliminate α from Eq. (A-23b) by first solving it for α, then writing $\dot{\alpha}$ in terms of β, $\dot{\beta}$ in Eq. (A-23a), and finally eliminating $\dot{\alpha}$ from Eq. (A-23b) after differentiating it once with respect to time. The equation for $\beta(t)$ is

$$\ddot{\beta} + i\omega\dot{\beta} - \frac{\beta}{4}(2\omega\omega_0 - \omega_0^2 - \omega_1^2) = 0 \tag{A-24}$$

The solution of Eq. (A-24) that satisfies $\alpha(0) = 1$ and the normalization condition $|\alpha(t)|^2 + |\beta(t)|^2 = 1$ is

$$\beta(t) = \frac{i\omega_1 e^{-i(\omega/2)t}}{\sqrt{(\omega-\omega_0)^2+\omega_1^2}}\sin\left\{[(\omega-\omega_0)^2+\omega_1^2]^{1/2}\frac{t}{2}\right\} \tag{A-25}$$

The probability that the spin is in the down state is

$$|\beta(t)|^2 = \frac{\omega_1{}^2}{(\omega - \omega_0)^2 + \omega_1{}^2} \sin^2\left\{[(\omega - \omega_0)^2 + \omega_1{}^2]^{1/2} \frac{t}{2}\right\} \tag{A-26}$$

Equation (A-26) is the same as Eq. (1-22) after we make the identification

$$\sin^2 \theta = \frac{\omega_1{}^2}{(\omega - \omega_0)^2 + \omega_1{}^2}$$

The student who fills in the algebraic steps between Eq. (A-22) and Eq. (A-26) will greatly appreciate the labor-saving device represented by the classical motion of the spin in the rotating frame.

Index